Secure Fusion Estimation and Control for Cyber-Physical Systems Under Malicious Attacks

Bo Chen · Tongxiang Li · Wen-An Zhang · Li Yu

Secure Fusion Estimation and Control for Cyber-Physical Systems Under Malicious Attacks

Springer

Bo Chen
Department of Automation
Zhejiang University of Technology
Hangzhou, Zhejiang, China

Tongxiang Li
Department of Automation
Zhejiang University of Technology
Hangzhou, Zhejiang, China

Wen-An Zhang
Department of Automation
Zhejiang University of Technology
Hangzhou, Zhejiang, China

Li Yu
Department of Automation
Zhejiang University of Technology
Hangzhou, Zhejiang, China

ISBN 978-981-97-9395-2 ISBN 978-981-97-9396-9 (eBook)
https://doi.org/10.1007/978-981-97-9396-9

This Springer imprint is published by the registered company Springer Nature Singapore Pte Ltd.
The registered company address is: 152 Beach Road, #21-01/04 Gateway East, Singapore 189721, Singapore

Preface

The efficient integration of physical processes and cyber infrastructure with ubiquitous computation resources and communication capabilities greatly boosts the rapid development of cyber-physical systems (CPSs) in both theoretical investigations and engineering applications. CPSs typically employ distributed networks consisting of multiple embedded sensing and actuation units, working together as a team to gather, process, and exchange information. Recently, CPSs with distributed subsystems have created new applications with enormous societal impact and economic benefit to smart buildings, intelligent transportation, and so on. As one of the important issues in CPSs, real-time state estimation based on sensor measurements has recently attracted considerable research interests, because state estimates can provide a CPS with real-time monitoring and control capability. With the rapid development of network communication technology, especially wireless network technology and low-power wireless sensor network technology, communication networks have been introduced into the multi-sensor fusion estimation system. As a result, the information transmission between sensors and fusion center is carried out in the communication network rather than through dedicated lines, forming a networked multi-sensor fusion estimation system. This system has the advantages of less wiring, easy maintenance, strong scalability, high flexibility, and convenient information sharing, making the structure of distributed fusion estimation systems easy to implement. However, with the data transmitted in the unprotected wireless network, the vulnerability of CPSs brings new risks of potential safety hazards, which may cause great damage such as economical loss, facility damages, and even enormous loss of human lives. Therefore, it is critical to safeguard the security of CPS, and the security of CPS has attracted considerable interest in the last decade. In this case, multi-sensor information fusion estimation (MIFE) provides an attractive alternative to study secure estimation problems because MIFE can potentially improve estimation accuracy and enhance the reliability and robustness against attacks.

This book presents a series of fusion-based security measures for cyber-physical systems under malicious attacks, including attack detection, privacy protection, secure fusion estimation, and secure control. First, a review on the latest developments in the literature is presented in Chap. 1. To ensure the security of transmitted data, the distributed secure estimation against eavesdroppers with the finite bandwidth is studied in Chap. 2. In Chap. 3, the Paillier homomorphic encryption method is introduced in the distributed fusion estimation. Then, Chap. 4 introduces the method that utilizes multi-sensor fusion strategy to improve the speed of attack detection and warning, considering the quantization of residual signals. Secure fusion estimation methods against false data injection (FDI) attacks and denial-of-service (DoS) attacks are proposed in Chaps. 5–7. For dual-channel FDI attacks, Chap. 5 proposes the secure H_∞ fusion estimation for linear discrete-time CPSs based on the intermediate variable. For nonlinear CPSs, the distributed fusion estimation problem where FDI attacks are occurring at the actuator is addressed in Chap. 6. Furthermore, a novel networked nonlinear fusion estimation method against DoS attacks is proposed in Chap. 7. The secure control methods for DoS attacks, time-delay attacks, and FDI attacks are presented in Chaps. 8–11. Specifically, the secure control approach based on the event-triggered idea and predictive control theory is developed in Chap. 8 to actively defend the DoS attacks. In Chap. 9, a new switched system model with norm-bounded uncertainties is proposed to characterize the time-delay attack and then a packet-based control strategy is designed. The model-free reinforcement learning method is adopted in Chap. 10 for unknown CPSs under time-delay attacks. For secure control against FDI attacks, an unknown input observer and a slide mode control law are designed to attenuate the impact of attacks in Chap. 11.

Hangzhou, China
April 2024

Bo Chen
Tongxiang Li
Wen-An Zhang
Li Yu

Acknowledgements This work was supported in part by the National Natural Science Foundation of China under Grant 61973277 and 62073292, by the Joint Funds of the National Natural Science Foundation of China under Grant U24A20258, by the Zhejiang Provincial Natural Science Foundation of China under Grant LR20F030004, by the funding of Leading Innovative and Entrepreneur Team Introduction Program of Zhejiang under Grant 2023R01006, and by the Fundamental Research Funds for the Provincial Universities of Zhejiang under Grant RF-C2023007.

Contents

Acronyms

$\mathfrak{R}$	Space of real numbers
$\mathfrak{R}^n$	n-dimensional real Euclidean space
$\mathfrak{R}^{m\times n}$	Space of all $m \times n$ real matrices
$\mathbf{I}_n/\mathbf{I}$	Identity matrix with dimension n (or appropriate dimensions)
$\mathbf{0}_n/\mathbf{0}$	Zero matrix with dimension n (or appropriate dimensions)
$A > 0$	Symmetric positive definite
$A \geq 0$	Symmetric positive semi-definite
$A < 0$	Symmetric negative definite
$A \leq 0$	Symmetric negative semi-definite
A^{T}	Transpose of matrix A
A^{-1}	Inverse of matrix A
$\mathrm{col}\{x_i\}$	A column vector composed of elements x_i
$\mathrm{Var}(x)$	Variance of the random vector x
Rank(A)	Rank of matrix A
$\rho(A)$	Spectral radius of matrix A
$\sigma(X)$	Denotes the set of eigenvalues of the matrix X
$\lambda_{max}(A)$	Maximum eigenvalue of matrix A
$\lambda_{min}(A)$	Minimum eigenvalue of matrix A
Tr(A)	Trace of matrix A
$\mathrm{diag}\{\cdots\}$	Diagonal matrix
$\|\cdot\|$	Euclidean norm of a vector or an induced norm of a matrix
$\otimes$	Hadamard product
sup	Supremum
inf	Infimum
lim	Limit
lcm	Least Common Multiple
mod	Modulo
$\mathcal{L}_2[0, \infty)$	Space of square integrable functions on $[0, \infty)$
$l_2[0, \infty)$	Space of square summable infinite sequences on $[0, \infty)$
$\mathrm{Prob}\{X = x\}$	Probability of $X = x$
$\mathbb{E}\{x\}$	Expectation of x

$\triangleq$	Define
ADP	Adaptive dynamic programming
BRL	Bounded real lemma
CPS	Cyber-physical system
DFE	Distributed fusion estimation
DoS	Denial-of-service
FC	Fusion center
FDI	False data injection
IoT	Internet of things
KF	Kalman filter
LMI	Linear matrix inequality
LSE	Local state estimator
LTI	Linear time-invariant
MC	Monitoring center
MFE	Multi-sensor fusion estimation
MSE	Mean square error
QSMD	Quantitative signal measurement deviation
RL	Reinforcement learning

Chapter 1
Introduction

1.1 Security of CPSs

The rapid progress in computer, communication, and control technologies has brought a substantial revolution in the interaction and integration between the information space and the physical world [1]. The growing complexity is further magnified by the advent of transformative technologies such as the Internet of Things (IoT), big data analytics, and cloud computing. These advancements have significantly accelerated the integration of digital and physical systems, presenting new challenges and opportunities. Traditionally, embedded technologies rely on point-to-point solutions to address the diverse challenges associated with physical systems. However, in the evolving landscape of information and intelligent systems, this approach no longer fully satisfies the requirements for real-time responsiveness, controllability, and scalability. The integration of these systems now demands more advanced and adaptive solutions to handle the growing complexity and interconnectedness. In response to the deepening integration of industrialization and information technology, the concept of cyber-physical systems (CPSs) has emerged, garnering significant attention from both academia and industry [2–5]. CPSs represent a new generation of systems that combine computing, networking, and physical processes to enable seamless interaction between the digital and physical worlds. This novel paradigm offers higher adaptability, autonomy, efficiency, and functionality compared with traditional embedded systems. The development of CPSs has the potential to revolutionize various industries, including manufacturing, transportation, energy, healthcare, and agriculture, by enabling smarter and more efficient processes. Through real-time data acquisition, advanced analytics, and control algorithms, CPSs can optimize performance, enhance safety, reduce energy consumption, and bring other benefits. At present, CPSs have become the bottom support of most key infrastructures. In February 2006, the American Academy of Sciences issued the "American Competitiveness Plan", which clearly listed CPSs as an important research project. In April

B. Chen et al., *Secure Fusion Estimation and Control for Cyber-Physical Systems Under Malicious Attacks*, https://doi.org/10.1007/978-981-97-9396-9_1

2013, Germany officially launched "Industry 4.0" at the Hannover Messe. China government has also issued a new round of national development strategies such as "Made in China 2025" and "Industrial Internet" to promote the development of CPSs.

As CPSs continue to gain momentum and development, it is crucial for researchers, engineers, and policymakers to work collaboratively to address the challenges associated with their design, implementation, and integration. Ensuring the security, privacy, and reliability of these systems will be paramount as they become increasingly embedded in our everyday lives and critical infrastructures. The ubiquitous computing and communication resources make it possible for pervasive multilayer CPSs, but this also brings additional security vulnerabilities [6, 7]. Specifically, compared with traditional networked control systems, the security threats of CPSs are more serious due to the following reasons: 1) their attack surface is wider and more scalable [8]; 2) adversaries can cause severe disruptions of physical processes through malicious operations at different network layers [9]; 3) any successful attack on CPSs may lead to catastrophic system failures and incur unbearable losses, which in turn become the interest-driven motives of enemies in the real world [10]. Several security incidents in modern infrastructures further attest to these facts, such as the Ukraine power grid attack, and the Stuxnet worm attack on Iran's nuclear control system [11]. These security incidents demonstrate that attacks on CPSs, especially those initiated in the network domain, could result in enormous losses in economy or even human lives. Therefore, it has become more important than ever in both academia and industry to ensure the security of CPSs by developing new analysis techniques and design strategies.

Based on the attacker's objectives, cyber attacks on CPSs can be categorized into three types: attacks on availability, attacks on integrity, and attacks on confidentiality. Availability attacks are the most prevalent cyber attacks that aim to disrupt communication networks by rendering data and information unavailable. Typical examples of availability attacks include denial-of-service (DoS) attacks [12, 13], distributed DoS (DDoS) attacks [14–16], and time-delay attacks [17–19]. Integrity attacks may occur in sensors, actuators, communication networks, computing, and control centers because data and control commands can be forged under such attacks. There are various types of integrity attacks, including false data injection (FDI) attacks [20–22], sparse attacks [23–25] and replay attacks [26–28]. Confidentiality attacks can happen in any part of the system since attackers may target any system information, including eavesdropping attacks [29–31].

The inherent limitations of certain CPSs applications render the sensors and controllers within the system incapable of integrating conventional security measures like firewalls or antivirus software. As a result, contemporary CPSs security approaches predominantly rely on the system's information, structure, and data, utilizing well-designed security algorithms or control strategies to achieve CPSs security protection. Since the wireless communication is implemented by broadcasting, the transmitted data is easily intercepted by placing the receiving antenna in the area through which the radio broadcast signal passes. Therefore, the confidentiality becomes a basic requirement and fundamental issue for CPSs. Moreover, in the event that CPSs privacy protection measures prove to be ineffective and an malicious attack occurs, it becomes imperative to develop strategies to mitigate the attack and implement methods for system state restoration. Through an evaluation of existing literature from the vantage point of control theory and system engineering, four primary research directions for CPSs security can be identified: privacy protection [32–34] , attack detection [35–39], secure state estimation [40–44], and secure control [45–49]. Attack detection aims to identify potential malicious attacks in CPSs and respond promptly to the threat. Secure state estimation involves constant monitoring and assessment of system states to detect potential security risks and to implement proactive measures to prevent their occurrence. Secure control involves the application of various control strategies in CPSs to facilitate real-time control and security management, preventing attacks or system damage. These research directions operate in conjunction to form a comprehensive CPS security protection solution.

1.2 Privacy-Preserving Fusion Estimation for CPSs

Communication network is the key to implement information transmission in multi-sensor fusion estimation systems and serves as a link for internal and external information exchanges of the system. However, it is also one of the important risk sources for cyber security and is susceptible to malicious attacks. By utilizing the broadcast characteristic of wireless transmissions, attackers can easily eavesdrop on local estimation data packets transmitted over the channel. This can result in a significant amount of sensitive information leaked, and if the eavesdropper analyzes this information, the confidential information will be acquired by unauthorized parties. Secure state estimation against eavesdroppers aims to prevent such attacks by using techniques such as encryption, authentication, and secure communication protocols to ensure the confidentiality and integrity of the information being transmitted and processed in CPSs. Classical cryptographic techniques have been extensively studied in the field of information security [50, 51]. However, the limited computing resources at the sensor side make it challenging to use strong encryption for confidential communication. Weak encryption schemes are also prone to information leakage if eavesdroppers have sufficient computational capacity. As an alternative, recent research has explored secure estimation problems from the signal processing perspective [52–57], utilizing physical layer information and artificial noise (AN) to ensure the confidentiality. However, the problem of distributed secure fusion estimation against eavesdroppers with the finite bandwidth was not fully investigated. Under the dimensionality reduction fusion framework, a sensor only transmits partial components of local sensor estimate to the FC. Detailed results will be presented in Chap. 2. Homomorphic encryption, which was introduced in [58], is also an efficient approach that employs a pair of keys and homomorphisms to ensure computational privacy. This method prevents eavesdroppers from obtaining valid information when the transmission is encrypted with secret keys. In contrast to complex noise insertion [59], homomorphic encryption does not require additional energy consumption. Additionally, it does not impair the performance of legitimate users like noise contamination [60] or differential privacy [61]. Hence, this book studies the privacy-preserving distributed fusion estimation problem by utilizing homomorphic encryption in Chap. 3.

1.3 Fusion-Based Attack Detection for CPSs

Different attack detection methods have been developed to meet the stringent safety and robustness requirements, which are inspired by the existing fault diagnosis/tolerance literatures [62, 72]. Particularly, how to detect cyber-attacks on process control systems has been formulated as the fault diagnosis problem in [63]. Then, the resilient attack detection estimators were delicately constructed in [64] to detect the false data. Combining the smart grid system with graph theory, an observer-based algorithm was proposed in [65] to detect and isolate FDI attacks by using real-time measurements. Notice that the above FDI attack detection methods are used to judge whether there exist attack signals. However, their response speed may not satisfy the requirements of some practical systems such as unmanned vehicles or other critical infrastructures. Moreover, artificially-designed attacks may cause greater damage to systems, and how to shorten the alarm response of attack detection is of great significance. Unfortunately, few results are reported from the perspective of improving alarm response speed. It is well-known that when the FDI attack detector is designed based on estimation methods, the estimation performance and the accuracy of detection threshold are two important factors for the alarm response of detectors. At the same time, multi-sensor fusion estimation (MFE) studies how to best utilize useful information contained in multiple sets of data for the purpose of estimating a parameter or process [66], which has become a critical issue in information fusion, and various fusion methods have been developed (see [67–69] and the references therein). Since MFE can potentially improve estimation accuracy and enhance reliability against attacks [70, 71], it provides a new research idea to study the problem of shortening the alarm response by improving the accuracy of detection thresholds. Though several secure fusion estimation methods were proposed in [72] for CPSs under attacks, the authors assumed that the attacks can be completely detected and also did not focus on alarm response speed of attack detectors. In this book, the alarm response problem of FDI attack detection based on multi-sensor fusion strategy will be investigated, where attacks exist in the control input commands of CPSs. To ensure real-time anomaly detection, a group of sensors deployed by different positions send their messages to a monitoring center (MC) through bandwidth-limited communication networks. Detailed results will be presented in Chap. 4.

1.4 Secure Distributed Fusion Estimation for CPSs

If an unauthorized party intercepts the transmission data and intentionally destroys or forges it, the received data by the decision-maker may not fully or even falsely represent the on-site information, which can result in incorrect command or strategy making. To counter such FDI attacks, secure state estimation, which is leveraged to fetch useful information (e.g., system states and attack pattern) from the corrupted measurements, has been extensively studied and numerous algorithms have been developed. In recent years, widely discussed secure state estimation methods can be divided into the following three categories depending on the location of FDI attacks: 1) sensor attacks, 2) actuator attacks, and 3) joint actuator and sensor attacks. For switched sensor attacks, [73] formulated the secure estimation problem into the classical error correction problem and proved that the maximum number of sensor attacks can be corrected. Meanwhile, instead of sensor attacks, authors in [74] investigated the attack-resilient state estimation problem for CPSs with sparse actuator attacks and process noises. Furthermore, considering the scenario where actuator and sensor are compromised simultaneously, [75] proposed a novel secure Luenberger-like observer by constructing an augmented system where the attacks were seen as part of the augmented states. In practice, with the widespread implementation and development of CPSs, the sent measurements and received control messages in communication networks are both subjected to unprecedented vulnerability to malicious attacks. This book investigates the secure H_∞ fusion estimation problem for linear discrete-time CPSs under actuator and sensor attacks in Chap. 5. On the one hand, it should be pointed out that most secure estimation methods focused on linear systems. Limited number of results are available for nonlinear system, which is probably due to the difficulties in analyzing the stability of desired secure estimation filter. On the other hand, owing to multiple available sensors, multi-sensor fusion estimation technique has gained an ever-increasing interest due to its high estimation accuracy and robustness. Fusion estimation techniques combined information can obtain more specific inferences and better estimation accuracy from multiple sensors than from a single sensor [76]. Based on different structures, fusion estimation can be separated into two basic types: the centralized fusion and the distributed fusion [77]. Different networked fusion estimation methods have been developed based on the two basic structures [78–84]. The attack estimation accuracy was improved in [85] by introducing a compensation factor. In order to improve security for CPSs, it is imperative to develop fusion criterion to further enhance the estimation performance. This book will study the distributed fusion estimation problems for a class of nonlinear systems under unknown but bounded noises in Chap. 6.

Since the DoS attack can block communication channels [86], many researchers focus on the effect of DoS attack against the performance of networked estimation. Generally, most of the existing works on DoS attacks rely heavily on the quantitative analysis and distributions of DoS attacks. Meanwhile, the statistical property of noises is demanded to be known in advance, which is impractical in real-world systems. Moreover, most networked systems in the real world are nonlinear, but few results focused on the nonlinear fusion estimation under DoS attacks. In this case, the following two questions in fusion estimation for nonlinear networked systems are still challenging: (1) How to compensate for the performance loss caused by DoS attacks on nonlinear systems? (2) How to design an algorithm to minimize the estimation error without any statistical information of noises or distributions of DoS attacks? Though a similar problem was considered in [87], the previous work focused on the linear systems. Motivated by these considerations, this book will propose a novel networked nonlinear fusion estimation method against DoS attacks in Chap. 7.

1.5 Secure Control for CPSs

Regarding secure control for CPSs against cyber attacks, there exist two major research directions: attack-tolerant control and attack-compensating control. The former focuses on developing effective control strategies and laws to tolerate unexpected anomalies induced by attacks, while the latter is concerned with designing compensation schemes that can prevent severe degradation or instability of system performance and stability. To achieve this, reliable attack detection mechanisms need to be established to accurately identify and locate the occurrence of network attacks. Most of the existing approaches assume that DoS attacks only occur in controller-to-actuator (C-A) channel or sensor-to-controller (S-C) channel [88–90], or occur in two channels synchronously [91, 92]. In fact, there is a more general scenario where the DoS attacks existing in two channels may be asynchronous. In the case of asynchronous DoS attacks, a small attack frequency may cause a long-time failure of control signals update compared with synchronous DoS attacks. Hence, the asynchronous DoS attacks are more difficult to be addressed [93]. Besides, in the most existing approaches, the actuator usually apply a "sample-and-hold control" strategy or "zero-input control" strategy when the actuator does not receive the control signals. In this case, the control performance of CPSs will be reduced because the control signal cannot be updated timely. Hence, the secure control approach defending DoS attacks will be introduced in Chap. 8. In terms of time-delaying attacks, some existing works have studied the time-varying delay problems in networked systems [94–96], which can be roughly categorized as follows: i) The stochastic system approach [97–100], where the delays are assumed to be random and subject to a certain distribution law; ii) The robust control approach [101–105], where the delays are transformed into parameter uncertainties due to the boundedness of delays; iii) The switched system approach [106], where the control system under time-varying delay is modeled as a switched system by using a special time-driven actuator which reads the buffer during a period T_0 no longer than the sampling period T (i.e., $T_0 = T/N$ for some positive integer N). However, the aforementioned methods cannot directly deal with the problem of attack-induced delays because the characteristics of attack-induced delays are different from those of time-varying delays. Specifically, it is difficult to justify that the time-delay attack obeys a certain probability distribution. A switched system approach with a mode-dependent controller will be proposed in Chap. 9 and a model-free RL method will be introduced in detail in Chap. 10. As being a robust and effective nonlinear control strategy, sliding mode control (SMC) is recognized as one of the most competent tools in dealing with uncertain systems owing to its robustness. Hence, SMC has been introduced for the purpose of ensuring the security of CPSs [107]. Specifically, M. Li et al. proposed an adaptive SMC law with a discontinuous input term to defend actuator attacks [108]. An SMC method was put forward by [109] to ensure that CPSs are stochastic finite-time bounded under random injection attacks. However, the aforementioned studies are under the same assumption, i.e., the system states can be directly measured. Nevertheless, this strict requirement is normally difficult to be satisfied in reality. Therefore, under more general assumptions, this book studies the control of CPSs under FDI attacks. Detailed results will be presented in Chap. 11.

1.6 Book Organization

The main purpose of this book is to introduce the latest research progresses in the security protection of CPSs. The materials used in the book are mainly based on the authors' research results. The organization of the book is as follows.

Chapter 1 reviews the research background and the latest progress in the security protection of CPSs.

Chapter 2 studies the distributed dimensionality reduction fusion estimation problem for cyber-physical systems with limited bandwidth in the presence of eavesdroppers. Since wireless communication is implemented by broadcasting, the eavesdroppers can collude to collect the data through other communication networks. To protect data privacy, based on the physical processes and the local estimation error covariance (EEC) matrix, an insertion method of artificial noise (AN) is developed such that only eavesdroppers' fusion EEC becomes worse. Meanwhile, the fusion center needs to decode the received signal due to the noise interference, while the successful decoding probability varies with signal to noise ratio. Subsequently, some criteria for the selection probabilities and the successful decoding probabilities are given to guarantee the effectiveness of the AN insertion strategy. Moreover, a sufficient condition of the designed AN power is derived to guarantee the confidentiality.

Chapter 3 investigates the privacy-preserving problem for the distributed fusion estimation scheme. When the legitimate user wants to obtain consistent information from multiple sensors, it always employs a fusion center (FC) to gather local data and compute distributed fusion estimates (DFEs). Due to the existence of the potential eavesdropper and the data exchanged among sensors, FC and user imperatively require privacy preservation. Hence, we propose a distributed confidentiality fusion structure against eavesdropper by using Paillier homomorphic encryption approach. In this case, FC cannot acquire real values of local state estimates, while it only helps calculate encrypted DFEs. Then, the legitimate user can successfully obtain the true values of DFEs according to the encrypted information and secret keys, which is based on the homomorphism of encryption.

Chapter 4 studies the alarm response problem of attack detection for linear and nonlinear CPSs under false data injection attacks. Notice that the faster the alarm response of the attack detection is, the less the system performance loss is. In this case, the distributed fusion strategy, which can potentially provide higher accuracy and reliability, is introduced for the first time to improve the speed of alarm response. Meanwhile, to ensure real-time anomaly detection, a group of sensors deployed by different positions send their messages to a monitoring center through communication networks. Due to the limited bandwidth, multiple finite-level logarithmic quantizers are used to reduce the size of data packages containing residual messages. Then, an optimal weighting fusion criterion is designed by establishing convex optimization problems such that the residual evaluation threshold is more accurate in the presence of quantization errors.

Chapter 6 studies the secure estimation problem for CPSs where the control input is falsified by the false data injection attack. Firstly, novel optimization objectives are constructed and attack estimation is performed from real and expected measurement outputs, where the estimator gains are jointly derived through solving quadratic problems recursively. Subsequently, the exponential stability of the proposed estimator is derived if the nonlinear observability rank condition is satisfied and an introduced factor is given properly. Moreover, to enhance estimation performance, a distributed fusion criterion is designed and weighting matrices are obtained without any statistical information.

Chapter 7 studies the distributed fusion estimation problem for nonlinear systems under DoS attacks, where it is not required to know the statistical property of DoS attacks and the bounds of system noises in advance. A compensation model is constructed to defend against DoS attacks in order to reduce the performance loss. Following the idea of bounded recursive optimization, two convex optimization problems, which are established in terms of linear matrix inequalities, are designed to obtain the local optimal estimator gains and the optimal weighting matrices by minimizing the upper bounds of estimation errors at each time.

Chapter 8 investigates the active secure control problem for CPSs under synchronous DoS attacks, where DoS attacks exist in both the S-C channel and the C-A channel. Due to the cost constraints of attacks, it is reasonable to consider that the number of maximum continuous DoS attacks in both the S-C and the C-A channels is bounded. Then, to defend the two-channel DoS attacks, an active secure control strategy that makes full use of the unattacked intervals is designed to ensure that the control inputs are updated timely in each period. Meanwhile, a secure controller that contains both the current and future control inputs is designed. Under the active secure control strategy and the secure controller, the addressed CPS under two-channel DoS attacks can be asymptotically stable without losing the control performance.

Chapter 9 investigates the modeling and stabilization problem for CPSs under time-delay attacks. First of all, the attack-induced delays are divided into multiple equilong subintervals and each subinterval is separated into a nominal part and an uncertain part. Then, the system is considered to dwell in different subsystems when the attack-induced delays fall into different subintervals. In this way, the CPS is modeled as a discrete-time switched system with norm-bounded uncertainties. Moreover, the mode-dependent controller is designed to defend the time-delay attacks and guarantee the exponential stability of the closed-loop CPS, which is switched according to the attack-induced delays and is implemented combined with the packet-based control strategy.

Chapter 10 presents an output regulation method for unknown CPSs under time-delay attacks in both the S-C channel and the C-A channel. The proposed approach is designed by using control inputs and tracking errors which are accessible data. Reinforcement learning is leveraged to update the control gains in real-time by using policy or value iterations. A thorough stability analysis is conducted and it is found that the proposed controller can sustain the convergence and asymptotic stability even when two channels are attacked.

Chapter 11 is concerned with the control of CPSs in the presence of malicious FDI attacks on actuators. The FDI attacks on actuators may result in faults of actuators or even the instability of CPSs. To tackle this problem, an unknown input observer (UIO) is proposed to estimate the system states and attack signals. For the aim of suppressing the impact of FDI attacks, a discrete-time sliding mode control (DSMC) algorithm is correspondingly put forward, where its reaching law is constructed based on the n-th order difference of the estimation of attack signals.

References

1. Guan XP, Yang B, Chen C et al (2016) A comprehensive overview of cyber-physical systems: from perspective of feedback system. IEEE/CAA J. Automatica Sinica 3(1):1–14
2. Li Y, Lv C, Zhang J et al (2018) High-precision modulation of a safety-critical cyber-physical system: control synthesis and experimental validation. IEEE/ASME Trans Mech 23(6):2599–2608
3. Rana MM, Li L, Su SW (2017) Cyber attack protection and control of microgrids. IEEE/CAA J Automat Sinica 5(2):602–609
4. Cao X, Cheng P, Chen J et al (2014) Cognitive radio based state estimation in cyber-physical systems. IEEE J Select Areas Commun 32(3):489–502
5. Mitchell R, Chen R (2013) Effect of intrusion detection and response on reliability of cyber physical systems. IEEE Trans Reliabil 62(1):199–210
6. Ge X, Han QL, Zhong M et al (2019) Distributed Krein space-based attack detection over sensor networks under deception attacks. Automatica 109:108557
7. Yuan H, Xia Y, Yang H (2020) Resilient state estimation of cyber-physical system with multichannel transmission under DoS attack. IEEE Trans Syst Man Cybernet: Syst 51(11):6926–6937
8. Ding D, Han QL, Ge X et al (2020) Secure state estimation and control of cyber-physical systems: a survey. IEEE Trans Syst Man, and Cybernet: Syst 51(1):176–190
9. De Persis C, Tesi P (2015) Input-to-state stabilizing control under denial-of-service. IEEE Trans Autom Control 60(11):2930–2944
10. Zhou J, Chen B, Yu L (2020) Intermediate-variable-based estimation for FDI attacks in cyber-physical systems. IEEE Trans Circ Syst II: Express Briefs 67(11):2762–2766
11. Cetinkaya A, Ishii H, Hayakawa T (2019) An overview on denial-of-service attacks in control systems: attack models and security analyses. Entropy 21(2):210
12. Kim J, Kim J, Kim H et al (2020) CNN-based network intrusion detection against denial-of-service attacks. Electron 9(6):916

13. Zhang H, Cheng P, Shi L et al (2015) Optimal denial-of-service attack scheduling with energy constraint. IEEE Trans Automat Cont 60(11):3023–3028
14. Kaur P, Kumar M, Bhandari A (2017) A review of detection approaches for distributed denial of service attacks. Syst Sci Cont Eng 5(1):301–320
15. Salim MM, Rathore S, Park JH (2020) Distributed denial of service attacks and its defenses in IoT: a survey. J Supercomput 76:5320–5363
16. Mahmoud MS, Hamdan MM, Baroudi UA (2020) Secure control of cyber physical systems subject to stochastic distributed DoS and deception attacks. Int J Syst Sci 51(9):1653–1668
17. Bianchin G, Pasqualetti F. (2018) Time-delay attacks in network systems. Cyber-Phys Syst Security, 157–174
18. Sargolzaei A, Yen KK, Abdelghani MN (2015) Preventing time-delay switch attack on load frequency control in distributed power systems. IEEE Trans Smart Grid 7(2):1176–1185
19. Xiahou KS, Liu Y, Wu QH (2020) Robust load frequency control of power systems against random time-delay attacks. IEEE Trans Smart Grid 12(1):909–911
20. Qi W, Hou Y, Zong G et al (2021) Finite-time event-triggered control for semi-Markovian switching cyber-physical systems with FDI attacks and applications. IEEE Trans Circ Syst I: Regular Papers 68(6):2665–2674
21. Chen Y, Qi D, Dong H et al (2020) A FDI attack-resilient distributed secondary control strategy for islanded microgrids. IEEE Trans Smart Grid 12(3):1929–1938
22. Zhang TY, Ye D (2020) False data injection attacks with complete stealthiness in cyber-physical systems: a self-generated approach. Automatica 120:109117
23. Shoukry Y, Tabuada P (2015) Event-triggered state observers for sparse sensor noise/attacks. IEEE Trans Automatic Control 61(8):2079–2091
24. An L, Yang GH (2017) Secure state estimation against sparse sensor attacks with adaptive switching mechanism. IEEE Trans Autom Cont 63(8):2596–2603
25. Hao J, Piechocki RJ, Kaleshi D et al (2015) Sparse malicious false data injection attacks and defense mechanisms in smart grids. IEEE Trans Ind Inf 11(5):1–12
26. Porter M, Hespanhol P, Aswani A et al (2020) Detecting generalized replay attacks via time-varying dynamic watermarking. IEEE Trans Autom Cont 66(8):3502–3517
27. Zhai L, Vamvoudakis KG (2021) A data-based private learning framework for enhanced security against replay attacks in cyber-physical systems. Int J Robust Nonlinear Cont 31(6):1817–1833
28. Elsaeidy AA, Jagannath N, Sanchis AG et al (2020) Replay attack detection in smart cities using deep learning. IEEE Access 8:137825–137837
29. Li X, Xu J, Dai HN et al (2015) On modeling eavesdropping attacks in wireless networks. J Comput Sci 11:196–204
30. Yu J, Lu L, Chen Y et al (2019) An indirect eavesdropping attack of keystrokes on touch screen through acoustic sensing. IEEE Trans Mob Comput 20(2):337–351
31. Yang W, Zheng Z, Chen G et al (2019) Security analysis of a distributed networked system under eavesdropping attacks. IEEE Trans Circ Syst II: Express Briefs 67(7):1254–1258
32. Feng Q, He D, Zeadally S et al (2019) A survey on privacy protection in blockchain system. J Netw Comput Appl 126:45–58
33. Sun PJ (2020) Security and privacy protection in cloud computing: discussions and challenges. J Netw Comput Appl 160:102642
34. Zhang D (2018/) Big data security and privacy protection. In: 8th international conference on management and computer science (ICMCS 2018). Atlantis Press, pp 275–278
35. Ao W, Song Y, Wen C (2016) Adaptive cyber-physical system attack detection and reconstruction with application to power systems. IET Cont Theo Appl 10(12):1458–1468
36. Chorppath AK, Alpcan T, Boche H (2015) Bayesian mechanisms and detection methods for wireless network with malicious users. IEEE Trans Mob Comput 15(10):2452–2465
37. Huang Y, Tang J, Cheng Y et al (2014) Real-time detection of false data injection in smart grid networks: an adaptive CUSUM method and analysis. IEEE Syst J 10(2):532–543
38. Kailkhura B, Han YS, Brahma S et al (2015) Distributed Bayesian detection in the presence of Byzantine data. IEEE Trans Sig Proc 63(19):5250–5263

39. Pasqualetti F, Dorfler F, Bullo F (2013) Attack detection and identification in cyber-physical systems. IEEE Trans Autom Cont 58(11):2715–2729
40. Anubi OM, Konstantinou C (2019) Enhanced resilient state estimation using data-driven auxiliary models. IEEE Trans Ind Inf 16(1):639–647
41. Chen Y, Kar S, Moura JMF (2018) Resilient distributed estimation: aensor attacks. IEEE Trans Autom Cont 64(9):3772–3779
42. Chin WL, Lee CH, Jiang T (2019) Blind false data attacks against AC state estimation based on geometric approach in smart grid communications. IEEE Trans Smart Grid 9(6):6298–6306
43. Fawzi H, Tabuada P, Diggavi S (2014) Secure estimation and control for cyber-physical systems under adversarial attacks. IEEE Trans Autom Cont 59(6):1454–1467
44. Ge X, Han QL, Wang Z (2017) A dynamic event-triggered transmission scheme for distributed set-membership estimation over wireless sensor networks. IEEE Trans Cybernet 49(1):171–183
45. An L, Yang GH (2018) LQ secure control for cyber-physical systems against sparse sensor and actuator attacks. IEEE Trans Cont Netw Syst 6(2):833–841
46. Ashok A, Govindarasu M, Wang J (2017) Cyber-physical attack-resilient wide-area monitoring, protection, and control for the power grid. Proc IEEE 105(7):1389–1407
47. Dibaji SM, Pirani M, Flamholz DB et al (2019) A systems and control perspective of CPS security. Ann Rev Cont 47:394–411
48. Ding D, Wang Z, Ho DWC et al (2016) Observer-based event-triggering consensus control for multiagent systems with lossy sensors and cyber-attacks. IEEE Trans Cybernet 47(8):1936–1947
49. Yang H, Xu H, Xia Y et al (2020) Stability analysis on networked control systems under double attacks with predictive control. Int J Robust Nonlinear Cont 30(4):1549–1563
50. Shannon CE (1949) Communication theory of secrecy systems. The Bell Syst Tech J 28(4):656–715
51. William S. (2006) Cryptography and network security: principles and practices. Pearson Education India
52. Liao WC, Chang TH, Chi Ma WK CY (2011) QoS-based transmit beamforming in the presence of eavesdroppers: an optimized artificial-noise-aided approach. IEEE Trans Signal Proc 59(3):1202–1216
53. Leong AS, Quevedo DE, Dolz D, Dey S (2017) Remote state estimation over packet dropping links in the presence of an eavesdropper. In arXiv [cs.SY]
54. Tsiamis A, Gatsis K, Pappas GJ (2017) State estimation with secrecy against eavesdroppers. IFAC-PapersOnLine 50(1):8385–8392
55. Goel S, Negi R (2008) Guaranteeing secrecy using artificial noise. IEEE Trans Wireless Commun 7(6):2180–2189
56. Leong AS, Redder A, Quevedo DE, Dey S. (2018) On the use of artificial noise for secure state estimation in the presence of eavesdroppers. In: 2018 European control conference (ECC)
57. Xu D, Chen B, Yu L (2018) Secure fusion estimation against eavesdroppers. In: 2018 37th Chinese control conference (CCC)
58. Rivest RL, Adleman L, Dertouzos ML (1978) On data banks and privacy homomorphisms. Foundat Secure Comput 4(11):169–180
59. Xu D, Chen B, Yu L (2022) Energy-constrained confidentiality fusion estimation against eavesdroppers 69(2):624–628
60. Yan X, Zhang Y, Xu D, Chen B (2022) Distributed confidentiality fusion estimation against eavesdroppers. IEEE Trans Aerospace and Elect Syst 58(4):3633–3642
61. Dwork C, McSherry F, Nissim K, Smith A. (2006) Calibrating noise to sensitivity in private data analysis. In: Theory of cryptography conference, pp 265–284
62. He X, Wang Z, Ji YD et al (2011) Robust fault detection for networked systems with distributed sensors. IEEE Trans Aerospace Electron Syst 47:166–177
63. Hwang I, Kim S, Kim Y et al (2009) A survey of fault detection, isolation, and reconfiguration methods. IEEE Trans Cont Syst Tech 18(3):636–653

64. Guan Y, Ge X (2017) Distributed attack detection and secure estimation of networked cyber-physical systems against false data injection attacks and jamming attacks. IEEE Trans Sig Inf Proc Over Netw 4(1):48–59
65. Luo X, Yao Q, Wang X et al (2018) Observer-based cyber attack detection and isolation in smart grids. Int J Electr Power Energy Syst 101:127–138
66. Li XR, Zhu Y, Wang J, et al. (2003) Optimal linear estimation fusion. I. Unified fusion rules. IEEE Trans Inf Theo 49(9): 2192–2208
67. Chen B, Yu L, Zhang WA et al (2014) Distributed H_∞ fusion filtering with communication bandwidth constraints. Sig Proc 96:284–289
68. Noack B, Sijs J, Reinhardt M et al (2017) Decentralized data fusion with inverse covariance intersection. Automatica 79:35–41
69. Chen B, Hu G, Ho DWC et al (2018) A new approach to linear/nonlinear distributed fusion estimation problem. IEEE Trans Autom Cont 64(3):1301–1308
70. Chen B, Ho DWC, Hu G et al (2017) Secure fusion estimation for bandwidth constrained cyber-physical systems under replay attacks. IEEE Trans on cybernetics 48(6):1862–1876
71. Chen B, Ho DWC, Zhang WA et al (2017) Distributed dimensionality reduction fusion estimation for cyber-physical systems under DoS attacks. IEEE Trans Syst Man, Cybernet: Syst 49(2):455–468
72. Wang Z, Wang D, Shen B et al (2018) Centralized security-guaranteed filtering in multirate-sensor fusion under deception attacks. J Franklin Inst 355(1):406–420
73. Chang YH, Hu Q, Tomlin CJ (2018) Secure estimation based Kalman filter for cyber-physical systems against sensor attacks. Automatica 95:399–412
74. Xie CH, Yang GH (2017) Secure estimation for cyber-physical systems under adversarial actuator attacks. IET Cont Theo Appl 11(17):2939–2946
75. Lu AY, Yang GH (2018) Secure Luenberger-like observers for cyber-physical systems under sparse actuator and sensor attacks. Automatica 98:124–129
76. Hall DL, Llinas J (1997) An introduction to multisensor data fusion. Proc IEEE 85(1):6–23
77. Deng Z, Zhang P, Qi W, Gao Y, Liu J (2013) The accuracy comparison of multisensor covariance intersection fuser and three weighting fusers. Inf Fusion 14(2):177–185
78. Ma J, Sun S (2013) Centralized fusion estimators for multisensor systems with random sensor delays, multiple packet dropouts and uncertain observations. IEEE Sens J 13(4):1228–1235
79. Michelusi N, Mitra U (2013) Fusion center feedback for quasi-decentralized estimation in sensor networks. In: 2013 51st annual allerton conference on communication, control, and computing (Allerton), pp 1285–1291
80. Assa A, Janabi-Sharifi F (2015) A kalman filter-based framework for enhanced sensor fusion. IEEE Sens J 15(6):3281–3292
81. Yang X, Zhang W, Yu L, Xing K (2016) Multi-rate distributed fusion estimation for sensor network-based target tracking. IEEE Sens J 16(5):1233–1242
82. Gao Y, Deng Z (2019) Robust weighted fusion kalman estimators for networked multisensor mixed uncertain systems with random one-step sensor delays, uncertain-variance multiplicative, and additive white noises. IEEE Sens J 19(22):10935–10946
83. Gao S, Zhong Y, Li W (2011) Random weighting method for multisensor data fusion. IEEE Sens J 11(9):1955–1961
84. Wu Y, Li Y, Chen B (2018) Distributed sinr fusion estimation for a class of wireless networks. IEEE Trans Circ Syst II: Express Briefs 65(9):1264–1268
85. Weng P, Chen B, Dong S et al (2020) Fusion estimation for FDI sensor attacks in distributed systems. In: 2020 IEEE 16th international conference on control and automation (ICCA). IEEE, pp 260–265
86. Carl G, Kesidis G, Brooks RR et al (2006) Denial-of-service attack-detection techniques. IEEE Internet Comput. 10(1):82–89
87. Zhu D, Chen B, Yu L (2019) Distributed fusion estimation for linear time-varying systems under DoS attacks and bounded noises. In: 2019 Chinese Control Conference (CCC). IEEE, pp 5480–5485

88. Dolk VS, Tesi P, De Persis C et al (2016) Event-triggered control systems under denial-of-service attacks. IEEE Trans Cont Netw Syst 4(1):93–105
89. Feng S, Tesi P (2017) Resilient control under denial-of-service: robust design. Automatica 79:42–51
90. Sun YC, Yang GH (2018) Periodic event-triggered resilient control for cyber-physical systems under denial-of-service attacks. J Franklin Inst 355(13):5613–5631
91. Feng S, Tesi P (2017) Networked control systems under denial-of-service: Co-located vs. remote architectures. Syst Cont Lett 108:40–47
92. Lai S, Chen B, Li T et al (2018) Packet-based state feedback control under DoS attacks in cyber-physical systems. IEEE Trans Circ Syst II: Express Briefs 66(8):1421–1425
93. Sun YC, Yang GH (2018) Event-triggered resilient control for cyber-physical systems under asynchronous DoS attacks. Inf Scie 465:340–352
94. Zhang D, Shi P, Wang QG et al (2017) Analysis and synthesis of networked control systems: a survey of recent advances and challenges. ISA Trans 66:376–392
95. Park P, Ergen SC, Fischione C et al (2017) Wireless network design for control systems: a survey. IEEE Commun Surv Tutor 20(2):978–1013
96. Ge X, Yang F, Han QL (2017) Distributed networked control systems: a brief overview. Inf Sci 380:117–131
97. Nilsson J, Bernhardsson B, Wittenmark B (1998) Stochastic analysis and control of real-time systems with random time delays. Automatica 34(1):57–64
98. Huang D, Nguang SK (2008) State feedback control of uncertain networked control systems with random time delays. IEEE Trans Auto Cont 53(3):829–834
99. Abdollahi F, Khorasani K (2010) A decentralized markovian jump H_∞ control routing strategy for mobile multi-agent networked systems. IEEE Trans Cont Syst Tech 19(2):269–283
100. Qi Q, Zhang H (2016) Output feedback control and stabilization for networked control systems with packet losses. IEEE Trans Cybernet 47(8):2223–2234
101. Cloosterman MBG, Van de Wouw N, Heemels W et al (2009) Stability of networked control systems with uncertain time-varying delays. IEEE Trans Automat Cont 54(7):1575–1580
102. Zhang WA, Yu L (2011) BIBO stability and stabilization of networked control systems with short time-varying delays. Int J Robust Nonlinear Cont 21(3):295–308
103. Zhang Z, Wang Z, Chen X et al (2017) Improved results on BIBO stability and stabilization of networked control systems with short time-varying delays. J Franklin Inst 354(17):7699–7716
104. Zhang WA, Yu L (2010) A robust control approach to stabilization of networked control systems with short time-varying delays. Acta Automat Sinica 36(1):87–91
105. Zhang H, Zhang Z, Wang Z et al (2015) New results on stability and stabilization of networked control systems with short time-varying delay. IEEE Trans Cybernet 46(12):2772–2781
106. Zhang WA, Yu L et al (2008) A switched system approach to networked control systems with time-varying delays. In: 2008 27th Chinese control conference. IEEE, pp 424–427
107. Huang X, Zhai D, Dong J (2018) Adaptive integral sliding-mode control strategy of data-driven cyber-physical systems against a class of actuator attacks. IET Cont Theo Appl 12(10):1440–1447
108. Li M, Chen Y, Zhang Y et al (2022) Adaptive sliding-mode tracking control of networked control systems with false data injection attacks. Inf Sci 585:194–208
109. Cao Z, Niu Y, Song J (2019) Finite-time sliding-mode control of Markovian jump cyber-physical systems against randomly occurring injection attacks. IEEE Trans Automat Cont 65(3):1264–1271

Chapter 2
Secure Dimensionality Reduction Fusion Estimation Against Eavesdroppers

2.1 Introduction

Due to the openness and anonymity of the network, cyber-attack has become a critical issue in CPSs. Particularly, the eavesdropping attack is a kind of dangerous cyber-attack, which can overhear the transmission in channels. It will leak the key information to malicious third-parties, and may cause significant economic losses and life risk. Besides, the bandwidth of communication channels may be limited in practical situations. In other words, the local data at sensors cannot be completely sent to the FC at every instant.

There exist many privacy-preserving approaches. The complex noise injection is adopted in this chapter, which is one of the effective methods based on the physical-layer information. More concretely, it adds complex noises into transmitted signals based on the communication channel gains. Notice that it can achieve perfect privacy, which means that the eavesdropper cannot acquire any valid information from channels, because it cannot successfully decode the signals. Moreover, the dimensionality reduction method is applied in this chapter to satisfy the bandwidth constraint, where only certain components can be sent through communication channels at each time. Then, the legitimate user utilizes one-step prediction to compensate the performance loss generated by the dimensionality reduction.

2.2 Problem Formulation

2.2.1 System Model

In this chapter, we consider the following state-space model:

$$x(t+1) = Ax(t) + w(t), \tag{2.1}$$

B. Chen et al., *Secure Fusion Estimation and Control for Cyber-Physical Systems Under Malicious Attacks*,https://doi.org/10.1007/978-981-97-9396-9_2

$$y_i(t) = C_i x(t) + v_i(t) \quad (i = 1, 2, \ldots, L), \tag{2.2}$$

where $x(t) \in \Re^n$ stands for the system state, $y_i(t) \in \Re^{q_i}$ represents the measurement of sensor i. $w(t)$ and $v_i(t)$ are independent zero-mean white Gaussian noises with variances Q and R_i, respectively. L is the number of sensors. A and C_i are respective system and measurement matrices, while the pair $(A, Q^{1/2})$ is controllable. Here, the sensor i will estimate the local state by using the standard Kalman filter [4–6] with the measurement $y_i(t)$, i.e.,

$$\hat{x}_i(t) = G_{K_i}(t)A\hat{x}_i(t-1) + K_i(t)y_i(t) \,, \tag{2.3}$$

$$K_i(t) = P_i^-(t)C_i{}^T(C_i P_i^-(t)C_i^{\mathrm{T}} + R_i)^{-1}, \tag{2.4}$$

where

$$G_{K_i}(t) = \mathbf{I}_n - K_i(t)C_i. \tag{2.5}$$

The local estimation error covariance (EEC) can be recursively calculated by

$$P_i(t) = G_{K_i}(t)P_i^-(t), \tag{2.6}$$

$$P_i^-(t) = AP_i(t-1)A^{\mathrm{T}} + Q. \tag{2.7}$$

Moreover, according to (2.3), (2.4), (2.5), (2.6) (2.7), the error cross-covariance $P_{ij}(t)$ can be computed according to [7]:

$$P_{ij}(t) = [\mathbf{I}_n - K_i(t)C_i][AP_{ij}^s(t-1)A^{\mathrm{T}} + Q][\mathbf{I}_n - K_j(t)C_j]^{\mathrm{T}} \,. \tag{2.8}$$

As mentioned before, the communication bandwidth and sensor energy in CPSs are limited. Dimensionality reduction is a classical approach to satisfy the bandwidth constraints [6, 7]. After calculating the local estimate $\hat{x}_i(t)$, only r_i $(1 \le r_i < n)$ components of $\hat{x}_i(t)$ will be selected to be sent to the FC at time t, which is denoted by $\hat{x}_i^s(t)$. Such a mechanism is called "random transmission" as shown in Fig. 2.1. In this case, the reorganized state estimate (RSE) $\hat{x}_i^r(t)$ of the FC has Δ_i possible cases in the following set:

$$S_i(t) \triangleq \{H_1^i\hat{x}_i(t), \ldots, H_{h_i}^i\hat{x}_i(t), \ldots, H_{\Delta_i(t)}^i\hat{x}_i(t)\} \,, \tag{2.9}$$

where $\Delta_i(t) = (n(n-1)\cdots(n-r_i+1))/(r_i(r_i-1)\cdots 1)$, $h_i = 1, 2, \cdots, \Delta_i(t)$. $H_{h_i}^i$ stands for a diagonal matrix consisting of r_i elements being 1 and $n - r_i$ elements being 0 on diagonal. For brevity, the indicator functions are described by

$$\alpha_{h_i}^i(t) = \begin{cases} 1 & if\ \hat{x}_i^r(t) = H_{h_i}^i\hat{x}_i(t), \\ 0 & if\ \hat{x}_i^r(t) \neq H_{h_i}^i\hat{x}_i(t), \end{cases} \quad h_i \in \{1, 2, \cdots, \Delta_i(t)\}. \tag{2.10}$$

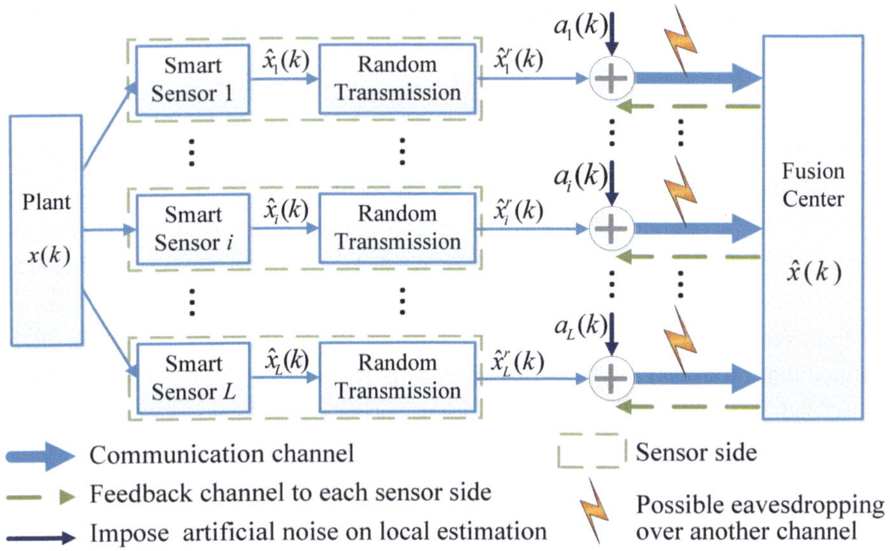

Fig. 2.1 Block diagram of the distributed dimensionality reduction fusion estimation

Since the energy is limited at certain level, sensor i may not send message to the FC, i.e., $\alpha_1^i(t) = \alpha_2^i(t) = \cdots \alpha_{\Delta_i(t)}^i(t) = 0$, which means that $\alpha_{h_i}^i(t)$ satisfies

$$\begin{cases} \alpha_{h_m}^i(t)\alpha_{h_n}^i(t) = 0 \quad (h_m \neq h_n), \\ \sum_{h_i=1}^{\Delta_i(t)} \alpha_{h_i}^i(t) \in \{0, 1\} \ (i = 1, 2, \ldots, L), \end{cases} \tag{2.11}$$

where $\sum_{h_i=1}^{\Delta_i(t)} \alpha_{h_i}^i(t) = 1$ means some components of the local estimates will be sent out, while $\sum_{h_i=1}^{\Delta_i(t)} \alpha_{h_i}^i(t) = 0$ indicates the sensor will not send any message. Then, the RSE $\hat{x}_i^r(t)$ can be expressed based on (2.9), (2.10)-(2.11):

$$\hat{x}_i^r(t) = H_i(t)\hat{x}_i(t) \ , \tag{2.12}$$

where $H_i(t) \triangleq \sum_{h_i=1}^{\Delta_i(t)} \alpha_{h_i}^i(t) H_{h_i}^i$ is a diagonal matrix satisfying

$$H_i(t) = \text{diag}\{\beta_1^{\text{i}}(\text{t}), \ldots, \beta_{\text{n}}^{\text{i}}(\text{t})\}, \tag{2.13}$$

with $\beta_j^i(t) \in \{0, 1\}$ $(j = 1, 2, \ldots, n)$. According to (2.10) and (2.11), one has $\sum_{j=1}^{n} \beta_j^i(t) \in \{0, r_i\}$ $(i = 1, \ldots, L)$. In order to reduce the dimensionality, the sensor will randomly send the component of $\hat{x}_i(t)$ to the FC. The matrix $H_i(t)$ can be reconstructed at the FC based on the identification of RSE [2]. Then, the detailed transmission can be known by $\alpha_{h_i}^i(t)$ according to (2.9)–(2.12). The value of $\alpha_{h_i}^i(t)$

is 1 or 0, and the occurrence probabilities can be expressed by

$$\text{Prob}\{\alpha^i_{h_i}(t) = 1\} = \pi^i_{h_i}, \ \text{Prob}\{\alpha^i_{h_i}(t) = 0\} = 1 - \pi^i_{h_i} , \tag{2.14}$$

where $\{\alpha^i_{h_i}(t)\}$ $(h_i = 1, 2, \cdots, \Delta_i(t))$ obeys an independent and identical distribution. Moreover, the positive scalar $\pi^i_{h_i}$ in (2.14) satisfies

$$0 \leq \sum\nolimits_{h_i=1}^{\Delta_i(t)} \pi^i_{h_i} \leq 1 . \tag{2.15}$$

Moreover, by recalling (2.13), $\beta^i_j(t)$ satisfies the Bernoulli distribution. It is assumed that $\text{Prob}\{\beta^i_j(t) = 1\} = \beta^i_j$ and $\text{Prob}\{\beta^i_j(t) = 0\} = 1 - \beta^i_j$. The following relation can be derived with (2.14):

$$\mathbb{E}\{\sum\nolimits_{j=1}^{n} \beta^i_j(t)\} = r_i \sum\nolimits_{h_i=1}^{\Delta_i(t)} \pi^i_{h_i} , \tag{2.16}$$

where $\beta^i_j(t) = 0$ means that the j-th component of $\hat{x}_i(t)$ will not be sent to the FC. According to the RSE in (2.12), the compensating state estimate (CSE) of $x(t)$ can be calculated as

$$\hat{x}^c_i(t) = H_i(t)\hat{x}_i(t) + [\mathbf{I}_n - H_i(t)]A\hat{x}(t-1) , \tag{2.17}$$

where $[\mathbf{I}_n - H_i(t)]A\hat{x}(t-1)$ is utilized to compensate for the components of $\hat{x}_i(t)$ that the FC does not receive.

2.2.2 *Distributed Dimensionality Reduction Fusion Estimation*

According to the $\hat{x}^c_i(t)$ in (2.17), the fusion estimate $\hat{x}(k)$ can be described as

$$\hat{x}(t) = \sum_{i=1}^{L} W_i(t)\hat{x}^c_i(t) , \tag{2.18}$$

where $W_1(t)$, $W_2(t)$, ..., $W_L(t)$ are the weighting matrices satisfying $\sum\limits_{i=1}^{L} W_i(t) = \mathbf{I}_n$. Thus, the optimal weighting matrices in the linear minimum variance sense can be computed as follows [8]:

$$[W_1(t), \ldots, W_L(t)] = [\mathbf{I}^{\mathrm{T}}_a(\Sigma(t))^{-1}\mathbf{I}_a]^{-1}\mathbf{I}^{\mathrm{T}}_a(\Sigma(t))^{-1} , \tag{2.19}$$

where $\Sigma(t) = \mathbb{E}\{[e_1^{\mathrm{T}}(t)\cdots e_L^{\mathrm{T}}(t)]^{\mathrm{T}}[e_1^{\mathrm{T}}(t)\cdots e_L^{\mathrm{T}}(t)]\}, \mathbf{I}_a = [\mathbf{I}_n, \mathbf{I}_n, \ldots, \mathbf{I}_n]^{\mathrm{T}} \in \Re^{nL\times n}$ and $e_i(t) = x(t) - \hat{x}_i^c(t)$. Meanwhile, the EEC of the fusion estimate can be described as

$$P(t) = [\mathbf{I}_a^{\mathrm{T}}(\Sigma(t))^{-1}\mathbf{I}_a]^{-1}. \tag{2.20}$$

Based on the analysis above, the pivotal issue for the optimal fusion criterion is to calculate $\Sigma(t)$.

Lemma 2.1 *[1]. Define two matrices $G \triangleq diag\{g_1, \ldots, g_n\}$, $Z \triangleq diag\{z_1, \ldots, z_n\}$, and a stochastic matrix $U \triangleq \begin{bmatrix} u_{11} & \cdots & u_{1n} \\ \vdots & \ddots & \vdots \\ u_{n1} & \cdots & u_{nn} \end{bmatrix}$. Let product $\odot$ for* G *and* Z *be defined by*

$$G \odot Z = \begin{bmatrix} g_1 z_1 & \cdots & g_1 z_n \\ \vdots & \ddots & \vdots \\ g_n z_1 & \cdots & g_n z_n \end{bmatrix}.$$

Then, the following relationship holds:

$$\mathrm{E}\{GUZ\} = (G \odot Z) \otimes \mathrm{E}\{U\},$$

where $\otimes$ represents the Hadamard product.

Define

$$\begin{cases} \hat{H} = \mathbb{E}\{diag\{H_1(t), \cdots, H_L(t)\}\}, \\ \bar{H} = \mathbb{E}\{[(\mathbf{I}_n - H_1(t))^{\mathrm{T}}, \cdots, (\mathbf{I}_n - H_L(t))^{\mathrm{T}}]^{\mathrm{T}}\}, \\ \Lambda_{ij} = \mathbb{E}\{H_i(t) \odot H_j(t)\}, \\ \Xi_{ij} = \mathbb{E}\{(\mathbf{I}_n - H_i(t)) \odot (\mathbf{I}_n - H_j(t))\}, \\ V_{ij} = \mathbb{E}\{H_i(t) \odot (\mathbf{I}_n - H_j(t))\}, \\ \hat{\Sigma}(t-1) = P(t-1)\mathbf{I}_a^{\mathrm{T}} \sum^{-1}(t-1), \\ \hat{A}_i(t-1) = G_{K_i}(t)A, \\ \hat{P}_i(t-1) = [P_{1i}^{\mathrm{T}}(t-1)\cdots P_{Li}^{T}(t-1)]^{\mathrm{T}}. \end{cases}$$

The covariance matrix $\Sigma_{ij}(t) = \mathbb{E}\{e_i(t)e_j^{\mathrm{T}}(t)\}$ can be expressed as [2]

$$\begin{aligned} \sum\nolimits_{ij} = {} & \Lambda_{ij} \otimes P_{ij}(t) + V_{ij} \otimes [\Phi_i^{\mathrm{T}} A^{\mathrm{T}} + G_{K_i}(t)Q] \\ & + V_{ji}^{T} \otimes [A\Phi_j(t) + QG_{K_j}^{\mathrm{T}}(t)] \\ & + \Xi_{ij} \otimes [AP(t-1)A^{\mathrm{T}} + Q], \end{aligned} \tag{2.21}$$

where

$$\begin{aligned}\Phi_i(t) = &\hat{\Sigma}(t-1)\bar{H}A\Phi_i(t-1)\hat{A}_i^{\mathrm{T}}(t-1) \\ &+\hat{\Sigma}(t-1)\hat{H}\hat{P}_i(t-1)\hat{A}_i^{\mathrm{T}}(t-1) \\ &+\hat{\Sigma}(t-1)\bar{H}QG_{K_i}^{T}(t)\hat{A}_i^{\mathrm{T}}(t-1)\,.\end{aligned} \tag{2.22}$$

Then, the weighting matrix $W_i(t)$ and EEC matrix $\Sigma_{ij}(t)$ can be derived by substituting (2.21) into (2.19) and (2.20). Thus, the optimal fusion estimate $\hat{x}(k)$ can be computed with (2.18).

Remark 2.1 According to (2.18), (2.19), (2.20), (2.21) (2.22), the computation of $P(t)$ is dependent of all the parameters, such as the noise characteristics and the probabilities of binary variables $\alpha_{h_i}^{i}(t)$. Moreover, the computation of matrix $P(t)$ has nothing to do with the measurements of system dynamics. Hence, the EEC can be calculated offline since the parameters are known in prior.

2.2.3 *Problem of Interest*

Definition 2.1 [3]. A mechanism satisfies perfect expected secrecy if and only if the following conditions hold:

$$\lim_{t\to\infty}\sup Tr\mathbb{E}\{P(t)\} < \infty\,, \tag{2.23}$$

$$\lim_{t\to\infty} Tr\mathbb{E}\{P^{e}(t)\} = \infty\,, \tag{2.24}$$

where $P^{e}(t)$ stands for the EEC of the eavesdropper.

Consequently, this chapter will consider the following two problems:

- For the fusion estimation under dimensionality reduction, the first aim is to design effective fusion criteria to achieve the perfect expected secrecy.
- Under the constraints of bandwidth and energy, the second aim is to conduct noise-energy-dependent and probability-dependent conditions to guarantee the effectiveness of the proposed secure estimator.

2.3 Main Result

To protect the data privacy, an artificial noise (AN) injecting method is adopted in this chapter, and the energy condition of AN will be conducted to achieve the perfect expected secrecy.

2.3.1 Artificial Noise Injecting Approach

As mentioned before, AN is exploited to preserve the privacy of transmission. It is assumed that the sensor will transmit the certain component $\hat{x}_i^s(t)$ to the FC. In this case, a variable $a_i(t)$ will be inserted into $\hat{x}_i^s(t)$ before transmission; otherwise, there is no need to inject $a_i(t)$. Then, the finally transmitted signal can be described as follows:

$$z_i(t) = \hat{x}_i^s(t) + a_i(t) \ , \tag{2.25}$$

where $a_i(t)$ represents the AN vector. Obviously, the transmitted signal $\hat{x}_i^s(t)$ will be perturbed by $a_i(t)$. Since the eavesdropper cannot acquire all the system dynamics, the utilization of all the system information to design AN may be effective. However, it is difficult to employ a great deal of data to design an appropriate AN. By contrast, the fusion EEC $P_i(t)$ is suitable, which consists of all the information according to (2.6). Notice that $H_i(t)$ is a diagonal matrix that is determined by random transmission structure. Thus, the utilization of $H_i(t)$ can enhance the privacy level of AN approach.

Specifically, the diagonal elements of $P_i^s(t) \triangleq H_i(t)P_i(t)$ are considered to construct a vector $\Theta_i(t) \in \Re^{1\times r_i}$. Let $\mathrm{N}(\Theta_i(t))$ stand for the null space of $\Theta_i(t)$. Then, the AN $a_i(t)$ is designed to lie in the null space $\mathrm{N}(\Theta_i(t))$ and satisfy $\Theta_i(t)a_i(t) = 0$. Define $\Phi_i(t) = [\theta_i^1(t), \theta_i^2(t), \ldots, \theta_i^{r_i-1}(t)]$ as an orthonormal basis matrix of $\mathrm{N}(\Theta_i(t))$, and $\Phi_i(t) \in \Re^{r_i\times(r_i-1)}$ satisfies

$$\Phi_i^{\mathrm{T}}(t)\Phi_i(t) = \mathbf{I}_{r_i-1} \ . \tag{2.26}$$

In this case, the AN $a_i(t)$ can be determined by

$$a_i(t) = \Phi_i(t)\varsigma_i(t) \ , \tag{2.27}$$

where the components of $\varsigma_i(t) \in \Re^{r_i-1}$ are zero-mean white Gaussian noises with variance $\sigma_i^2(t)$. It should be pointed out that $\sigma_i^2(t)$ can be designed at the FC to ensure the effectiveness of the AN approach. Thus, the data received by the FC can be expressed by

$$z_i(t) = \hat{x}_i^s(t) + \Phi_i(t)\varsigma_i(t). \tag{2.28}$$

To eliminate the AN perturbation, the received data will be multiplied by $\Theta_i(t)$ for decoding, that is

$$z_i^u(t) = \Theta_i(t)z_i(t) = \Theta_i(t)\hat{x}_i^s(t) \ . \tag{2.29}$$

According to (2.29), it can be seen that the transmitted information $\hat{x}_i^s(t)$ will not be influenced by the AN $a_i(t)$.

Remark 2.2 $P_i(t)$ can be calculated by the sensor and the legitimate user based on the system matrices. At the same time, the random matrix can be reconstructed by the identification process of RSE in the FC, while $\Phi_i(t)$ can also be computed with local parameters. Moreover, according to (2.29), the transmitted packages will not be influenced by the added AN. Nonetheless, it is difficult for eavesdroppers to acquire $P_i(t)$. Although all these parameters are intercepted by the eavesdroppers, the precise transmitted signal still cannot be acquired due to the existence of the random matrix $H_i(t)$. Consequently, we can derive the suitable AN variance $\sigma_i^2(t)$ to satisfy the transmission privacy.

2.3.2 *Perfect Expected Secrecy Under Constrained Bandwidth and Energy*

Here, the suitable AN power will be determined to satisfy the perfect expected secrecy conditions (2.23) and (2.24). One key property of communication channels is the random disturbance, including complex noises, fading, and so on [9, 10]. After receiving the transmitted packets, FC will decode them while the successful decoding probability is related to the signal-to-noise ratio (SNR) [11]. Define $\delta_i(t)$ as the energy emitted by the sensor. Then, the SNR of the FC can be described as

$$\varepsilon_i(t) = \frac{\delta_i(t)G_i(t)}{\sum_{a_i(t)}}, \tag{2.30}$$

where $G_i(t)$ stands for the channel gain and $\sum_{a_i(t)}$ is the channel noise power. A binary variable $\gamma_i(t)$ is considered to denote whether the FC decodes the received message successfully, i.e. $\gamma_i(t) = 1$ means a successful decoding, $\gamma_i(t) = 0$ means a failed decoding, which indicates that the FC cannot obtain any valid data.

Additive White Gaussian Noise (AWGN) network is a fundamental model for interference channels. The successful decoding probability can be described by [12, 13]:

$$\text{Prob}(\gamma_i(t) = 1|\varepsilon_i(t)) = f(\varepsilon_i(t)) = [1 - \xi(\sqrt{2\varepsilon_i(t)})]^m , \tag{2.31}$$

where the Gauss ξ-function $\xi(x) = \int_x^{\infty} \frac{1}{\sqrt{2\pi}} \exp(-\frac{t^2}{2})dt$, and m is the packet length. Hence, according to (2.29) and (2.30), the SNR of the FC is ∞. Nevertheless, SNR of the eavesdropper can be calculated by the following form with (2.30):

$$\varepsilon_i^e(t) = \frac{\delta_i(t)G_i(t)}{\sigma_i^2(t)}, \quad i = 1, 2, \ldots, L. \tag{2.32}$$

Remark 2.3 According to (2.31), the successful decoding probability at the FC tends to 1 as $\varepsilon_i^e(t)$ goes to ∞. Therefore, the estimation performance of the legitimate user is always optimal. Due to the influence of AN, the SNR of the received packet by the eavesdropper is decreased based on (2.31) and (2.32), which results in a lower probability of successful decoding. Then, the EEC of eavesdropper will be proved to satisfy the perfect expected secrecy in (2.24).

For stable systems, the eavesdropper can always get a bounded EEC by prediction. Thus, the privacy protection for the stable system is more challenging, which is our future work. Instead, an unstable system is considered in this chapter, whose spectral radius satisfies $\rho(A) > 1$. According to (2.32), the SNR of the eavesdropper depends on $\delta_i(t)$, $G_i(t)$ and $\sigma_i^2(t)$. Hence, the perfect secrecy can be realized by determining these parameters and feeding them back to sensors.

Define $\rho_i \triangleq [\pi_1^i, \pi_2^i, \ldots, \pi_{\Delta_i}^i]^T$ $(i = 1, 2, \ldots, L)$, and $\gamma_i^e(t)$ denotes whether the received message can be successfully decoded for the eavesdropper with Prob$\{\gamma_i^e(t) = 1\} = \gamma_i^e$. According to (2.13), there exists a positive constant $\chi_j^i \in \mathfrak{R}^{1\times\Delta_i}$ such that

$$\beta_j^i = \chi_j^i \rho_i, \quad j = 1, 2, \ldots, n, \tag{2.33}$$

where an arbitrary element of χ_j^i is 1 or 0.

Theorem 2.1 *Consider an unstable system in (2.1). The encryption mechanism (2.28) is employed for secure transmission. The perfect expected secrecy for the proposed fusion estimator is achieved if the following conditions are simultaneously satisfied:*

(C1) For the L measurement equations defined in (2.2), at least one local subsystem i is observable.

(C2) There exists at least one set of selected transmission probabilities $\pi_{h_i}^i$, at every moment t, $h_i = 1, 2, \cdots, \Delta_i(t)$, satisfying

$$\lambda_{\max}\left\{A^{\mathrm{T}} M_i A\right\} < 1\ , \tag{2.34}$$

or

$$\begin{cases} \lambda_{\max}\left\{A^{\mathrm{T}} M_i A\right\} = 1 \\ \lambda_{\max}\left\{A^{\mathrm{T}} M_i A\right\} \neq \lambda_{\min}\left\{A^{\mathrm{T}} M_i A\right\} \end{cases}, \tag{2.35}$$

where $M_i = diag\{1 - \chi_1^i \rho_i, \ldots, 1 - \chi_n^i \rho_i\}$.

(C3) There exists a positive real number $\bar{\lambda} > 0$ such that each set of the selected transmission probabilities and the successful decoding probability at each time satisfy

$$Tr\left\{N_i^e \otimes P_{ii}(t) + M_i^e \otimes \left[AP^e(t-1)A^{\mathrm{T}} + Q\right] - \sum\nolimits_{ii}^{e}(t-1)\right\} > \bar{\lambda}, \tag{2.36}$$

where $M_i^e = diag\{(1 - \gamma_i^e \beta_1^i), \ldots, (1 - \gamma_i^e \beta_n^i)\}$, $N_i^e = diag\{\gamma_i^e \beta_1^i, \ldots, \gamma_i^e \beta_n^i\}$, and $\sum_{ii}^e(t-1)$ is the EEC of the eavesdropper.

Proof According to Remarks 2.2–2.3, the AN will not affect the performance of the legitimate user. Thus, the result (2.23) can be directly obtained based on (C1) and (C2) with some similar derivations as in [2], and the corresponding proof is omitted here.

Moreover, another perfect expected secrecy result (2.24) will be shown under the above conditions. For simplicity, all variables with superscripts are utilized to denote the corresponding variables of the eavesdropper in the following proof.

The covariance $\Sigma^e_{ii}(t)$ of the eavesdropper can be described by

$$\begin{aligned}\Sigma^e_{ii}(t) = {} & \Lambda^e_{ii} \otimes P_{ii}(t) + V^e_{ii} \otimes [\Phi_i^{\mathrm{T}}(t)A^{\mathrm{T}} + (\mathbf{I}_n - K_i(t)C_i)Q] \\ & +(V^e_{ii})^{\mathrm{T}} \otimes [A\Phi_i(t) + Q(\mathbf{I}_n - K_i(t)C_i)^{\mathrm{T}}] \\ & +\Xi^e_{ii} \otimes [AP^e(t-1)A^{\mathrm{T}} + Q]\,. \end{aligned} \tag{2.37}$$

Define

$$\begin{cases} m_1(t) = Tr(\Lambda^e_{ii} \otimes P_{ii}(t)), \\ m_2(t) = Tr(V^e_{ii} \otimes [\Phi_i^{\mathrm{T}}(t)A^{\mathrm{T}} + (\mathbf{I}_n - K_i(t)C_i)Q]), \\ m_3(t) = Tr(\Xi^e_{ii} \otimes [AP^e(t-1)A^{\mathrm{T}} + Q]). \end{cases} \tag{2.38}$$

Then, it can be concluded from (2.37) and (2.38) that the trace of $\Sigma^e_{ii}(t)$ is

$$Tr\{\Sigma^e_{ii}(t)\} = m_1(t) + 2m_2(t) + m_3(t). \tag{2.39}$$

Then, the following result can be deduced by considering the definition of Λ_{ij}:

$$\begin{aligned}\Lambda^e_{ii} &= E\left\{\gamma_i^e(t)\begin{bmatrix} \beta_1^i(t)\beta_1^i(t) & \cdots & \beta_1^i(t)\beta_n^i(t) \\ \vdots & \vdots & \vdots \\ \beta_n^i(t)\beta_1^j(t) & \cdots & \beta_n^i(t)\beta_n^i(t)\end{bmatrix}\right\} \\ &= \begin{bmatrix} \gamma_i^e\beta_1^i & \cdots & \gamma_i^e\beta_1^i\beta_n^i \\ \vdots & \vdots & \vdots \\ \gamma_i^e\beta_n^i\beta_1^i & \cdots & \gamma_i^e\beta_n^i \end{bmatrix}. \end{aligned} \tag{2.40}$$

Moreover, according to the trace operator, one has

$$m_1(t) = Tr(N_i^e \otimes P_{ii}(t))\,. \tag{2.41}$$

Similar to (2.40), we derive the following result with the definition of V_{ii}:

$$\begin{aligned}V^e_{ii} &= E\left\{\gamma_i^e(t)\begin{bmatrix} \beta_1^i(t)(1-\beta_1^i(t)) \cdots \beta_1^i(t)(1-\beta_n^i(t)) \\ \vdots \qquad\qquad \vdots \qquad\qquad \vdots \\ \beta_n^i(t)(1-\beta_1^i(t)) \cdots \beta_n^i(t)(1-\beta_n^i(t))\end{bmatrix}\right\} \\ &= \begin{bmatrix} 0 & \cdots & \gamma_i^e\beta_1^i(1-\beta_n^i) \\ \vdots & \vdots & \vdots \\ \gamma_i^e\beta_n^i(1-\beta_1^i) & \cdots & 0 \end{bmatrix}. \end{aligned} \tag{2.42}$$

By resorting to the properties of matrix trace operations, we have

$$m_2(t) = 0\,. \tag{2.43}$$

Moreover, based on the definition of Ξ_{ij}, Ξ^e_{ii} can be computed as

$$\Xi^e_{ii} = E\left\{\begin{bmatrix} (1-\gamma^e_i(t)\beta^i_1(t))(1-\gamma^e_i(t)\beta^i_1(t)) & \cdots & (1-\gamma^e_i(t)\beta^i_1(t))(1-\gamma^e_i(t)\beta^i_n(t)) \\ \vdots & \vdots & \vdots \\ (1-\gamma^e_i(t)\beta^i_n(t))(1-\gamma^e_i(t)\beta^i_1(t)) & \cdots & (1-\gamma^e_i(t)\beta^i_n(t))(1-\gamma^e_i(t)\beta^i_n(t)) \end{bmatrix}\right\}$$
$$= \begin{bmatrix} (1-\gamma^e_i\beta^i_1) & \cdots & (1-\gamma^e_i\beta^i_1)(1-\gamma^e_i\beta^i_n) \\ (1-\gamma^e_i\beta^i_n)(1-\gamma^e_i\beta^i_1) & \cdots & (1-\gamma^e_i\beta^i_n) \end{bmatrix}. \tag{2.44}$$

Then, the following result is derived with the property of the trace operator:

$$m_3(t) = Tr(M_i \otimes [AP(t-1)A^{\mathrm{T}} + Q]) . \tag{2.45}$$

By combining (2.39), (2.41), (2.43) and (2.45), $Tr\{\Sigma^e_{ii}(t)\}$ can be rewritten by

$$Tr\{\Sigma^e_{ii}(t)\} = Tr\{N^e_i \otimes P_{ii}(t) + M^e_i \otimes [AP(t-1)A^{\mathrm{T}} + Q]\} . \tag{2.46}$$

Hence, it follows from (C3) that there exists a positive number $k_i(t-1)$ greater than 1 for $t \geq N$:

$$\begin{aligned} Tr \sum\nolimits^e_{ii}(t) &\geq k_i(t-1) \cdot Tr(\sum\nolimits^e_{ii}(t-1)) \\ &\geq \cdots \\ &\geq \prod_{\tau=1}^{t-N} k_i(t-\tau) \cdot Tr(\sum\nolimits^e_{ii}(N)) . \end{aligned} \tag{2.47}$$

Based on (2.20) and (2.47), the EEC $P^e(t)$ of the eavesdropper is deduced as

$$\begin{aligned} Tr\{P^e(t)\} &= Tr(\mathbf{I}^{\mathrm{T}}_a(\sum\nolimits^e(t))^{-1}\mathbf{I}_a)^{-1} \\ &= Tr(\sum_{i=1}^{L}(\sum\nolimits^e_{ii}(t))^{-1})^{-1} \\ &\geq \tfrac{1}{L} Tr(\sum\nolimits^e_{ii}(t)), \qquad \text{for some } i . \end{aligned} \tag{2.48}$$

Moreover, by exploiting the limitation operation on both sides of (2.48), one has

$$\lim_{t\to\infty} Tr\{P^e(t)\} \geq \lim_{t\to\infty} \frac{1}{L}\prod_{\tau=1}^{t-N} k_i(t-\tau)\cdot Tr\{\Sigma^e_{ii}(N)\} \to \infty . \tag{2.49}$$

Finally, the condition (2.24) is obtained, i.e., $\lim_{t\to\infty} Tr\{\mathbb{E}\{P^e(t)\}\} = \infty$.

Remark 2.4 Note that the observability condition (C1) can be easily satisfied, because it is determined by system dynamics. As pointed out in [2], it is easy to give a set of random transmission probabilities to guarantee the convergence of EEC of the legitimate user. On the other hand, according to the proof of the theorem above, the EEC of the estimator of the eavesdropper will diverge to infinity if the condition (C1) and inequality (2.34), (2.35), (2.36) are satisfied. Since the conditions (C2) -

(C3) depend on the variables $P_{ii}(t)$, $P^e(t-1)$ and $\sum_{ii}^e(t-1)$, it is difficult to find an effective probability selecting range so that the EEC diverges. Based on (2.34)–(2.36), the conditions are independent of the measurements, which means that the transmission probability $\pi_{h_i}^i$ and the successful decoding probability γ_i^e can be calculated offline. Choosing $\pi_{h_i}^i$ and γ_i^e to satisfy both the conditions (C2) and (C3) will waste much time. Instead, probability $\pi_{h_i}^i$ can be firstly chosen for fitting (C2), and then select the decoding success probability γ_i^e to meet (C3). By combining (2.31), (2.32) and (2.36), the high energy of AN leads to a low probability of successful decoding, and therefore (C3) can be more easily satisfied. Then, the selection of AN power $\sigma_i^2(t)$ will be shown in Theorem 2.2.

Theorem 2.2 *Consider an unstable system, if the condition (2.36) is satisfied, the energy of AN should satisfy the following inequality:*

$$\sigma_i^2(t) > \frac{1}{\varepsilon_i^{e*}}\delta_i(t)G_i(t)\ , \tag{2.50}$$

where ε_i^{e*} is the SNR for the eavesdropper with respect to the probability of successful decoding γ_i^e.

Proof According to (2.31) and (2.32), the probability function monotonically increase with the SNR. Thus, the SNR of the eavesdropper should satisfy

$$\varepsilon_i^e(t) = \frac{\delta_i(t)G_i(t)}{\sigma_i^2(t)} < \varepsilon_i^{e*}\ . \tag{2.51}$$

Next, the result can be directly derived with the condition above.

Remark 2.5 From the theorem above, it can be concluded that the larger $\sigma_i^2(t)$ is, the more easily the inequality (2.36) is satisfied. It implies that the sensor needs to expend more energy. Notice that $\sigma_i^2(t)$ is required be transmitted to the sensor through the feedback channel. Nonetheless, if sensors have enough energy resources, sufficiently large energy of the AN $\sigma_i^2(t)$ can be chosen to satisfy the power condition (2.50). In this case, the feedback of the time-varying power $\sigma_i^2(t)$ is not required. Actually, the constant AN energy can achieve perfect encryption with minor energy, and the simulation result will be given in the sequel.

2.4 Simulation

To start with, the parameters for the systems (2.1)–(2.2) are set as follows:

$$A = \begin{bmatrix} 1.1 & 0.5 \\ 0 & 1 \end{bmatrix}, C_1 = C_2 = \begin{bmatrix} 1 & 0 \\ 0 & 1 \end{bmatrix}, Q = \begin{bmatrix} 0.5 & 0.25 \\ 0.25 & 1 \end{bmatrix},$$
$$R_1 = \text{diag}\{0.5, 0.1\}, \text{R}_2 = \text{diag}\{0.1, 0.3\}\ .$$

Based on the system above, we have rank$([C_i, C_iA]) = 2$, which means the condition (C1) is satisfied. Nevertheless, due to the limited bandwidth and energy, only one component of $\hat{x}_i(t)$ will be transmitted through the communication channels, i.e., $r_1 = r_2 = 1$. Then, according to (2.9), we can derive that

$$H_1^1(t) = H_1^2(t) = \begin{bmatrix} 1 & 0 \\ 0 & 0 \end{bmatrix}, H_2^1(t) = H_2^2(t) = \begin{bmatrix} 0 & 0 \\ 0 & 1 \end{bmatrix}. \quad (2.52)$$

Furthermore, considering the failure of decoding, the CSEs can be calculated by combining (2.13), (2.17) and (2.51):

$$\hat{x}_i^c(t) = \begin{bmatrix} \gamma_i^e(t)\beta_1^i(t) & 0 \\ 0 & \gamma_i^e(t)\beta_2^i(t) \end{bmatrix} \hat{x}_i(t) + \begin{bmatrix} 1-\gamma_i^e(t)\beta_1^i(t) & 0 \\ 0 & 1-\gamma_i^e(t)\beta_2^i(t) \end{bmatrix} A\hat{x}(t-1) . \quad (2.53)$$

where $\text{Prob}\{\beta_1^i(t) = 1\} = \beta_1^i$, $\text{Prob}\{\beta_2^i(t) = 1\} = \beta_2^i$, and $\text{Prob}\{\gamma_i^e(t) = 1\} = \gamma_i^e$. Here, β_1^1, β_1^2, β_2^1, and β_2^2 are all set as 0.5. Hence, we can deduce from (2.33) and (2.34) that

$$\lambda_{\max}\left\{A^{\mathrm{T}} M_i A\right\} = 0.8902 < 1 , \quad (2.54)$$

which indicates that condition (C2) holds.

Then, we assume that the transmission energy and channel gain are fixed with $\delta_i(t)$ and $G_i(t)$. The initial state is set as $x(0) = [0.15, 0.25]^T$, while the initial local estimate and fusion estimate are all chosen as $[0.1, 0.2]^T$. Other parameters are chosen as $P(0) = diag\{0.2, 0.3\}$, $\Phi_1(0) = \begin{bmatrix} 0.02 & 0.03 \\ 0.01 & 0.02 \end{bmatrix}$, and $\Phi_2(0) = \begin{bmatrix} 0.04 & 0.02 \\ 0.05 & 0.02 \end{bmatrix}$. Besides, the packet length is m=5.

First, the constant AN power is taken as $\sigma_i^2(t) = 0.9077$. Based on (2.31) and (2.32), the probability of successful packet decoding is $\text{Prob}\{\gamma_i^e(t) = 1\} = 0.7$. Moreover, it can be concluded from (2.36) that $M_i^e = diag\{0.65, 0.65\}$ and $N_i^e = diag\{0.35, 0.35\}$, and thus the condition (C3) is not satisfied. The estimation performance of local CSEs $\hat{x}_i^c(t)(i = 1, 2)$ and fusion estimate $\hat{x}^e(t)$ of the eavesdropper, and optimal estimate $\hat{x}(t)$ of the legitimate user are shown in Fig. 2.2. It shows that each local CSE performance is worse than that of the fusion estimation, which indicates that the eavesdropper can obtain a better performance by wiretapping more information from the sensors. However, the fusion estimation performance of the eavesdropper is worse than that of the optimal estimate of user, since there is no noise perturbation. On the other hand, owning to the high successful decoding rate, the boundedness of EEC of the eavesdropper is obtained based on the wiretapped data from two wireless networks.

Then, the AN power is chosen as $\sigma_i^2(t) = 5.6097$, and thus the probability of successful decoding becomes $\text{Prob}\{\gamma_i^e(t) = 1\} = 0.2$, which satisfies condition (C3). The estimation performance of the eavesdropper is shown in Fig. 2.3. If the proposed encryption mechanism is embedded in the communication data, it can be concluded that the traces of both local CSEs and fusion EEC of the eavesdropper grow unbounded. Besides, let us define

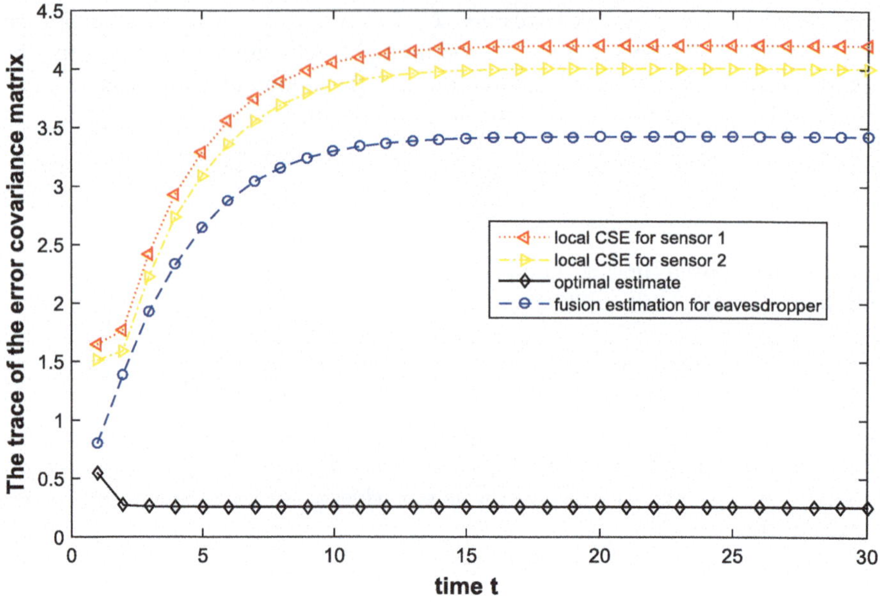

Fig. 2.2 The estimation performance with $\sigma_i^2(t) = 0.9077$

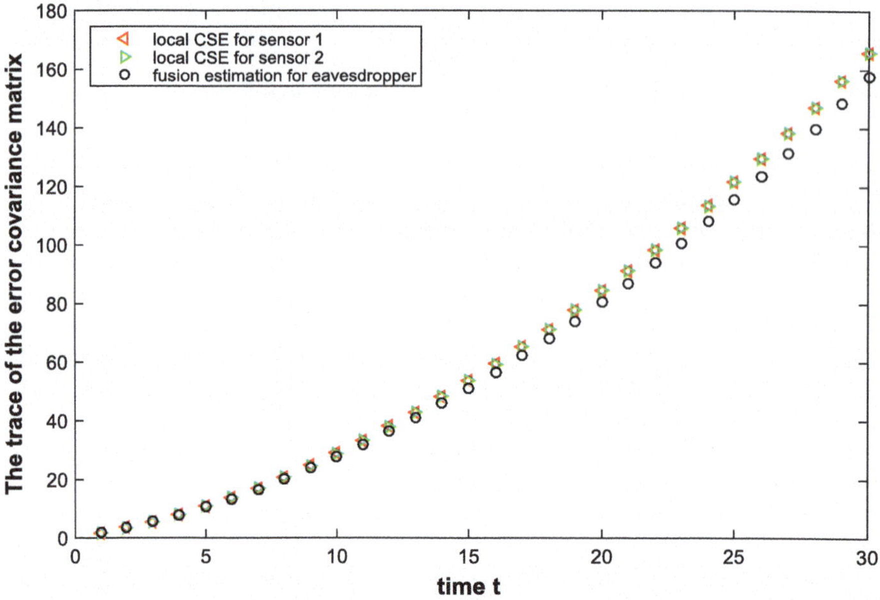

Fig. 2.3 The estimation performance for eavesdropper with $\sigma_i^2(t) = 5.6097$

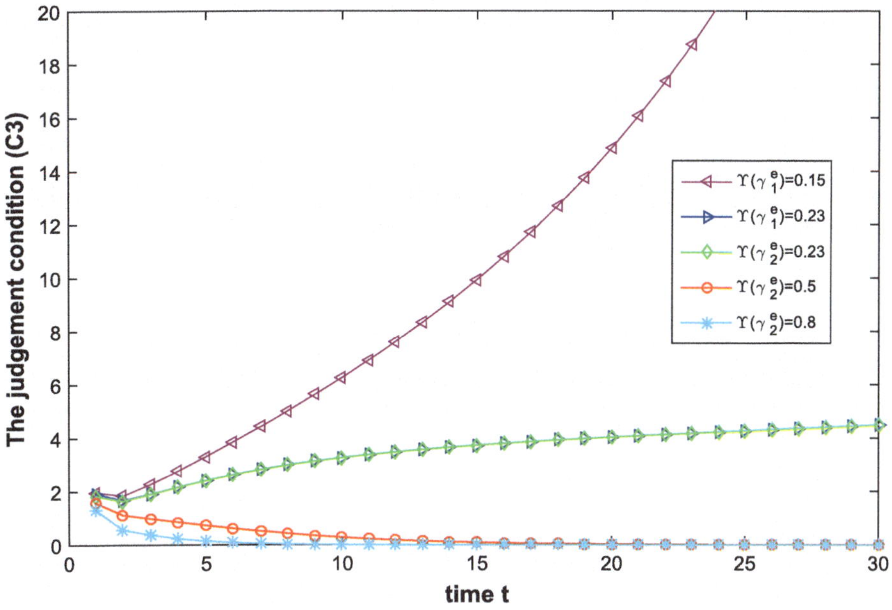

Fig. 2.4 The trajectories of $\Upsilon(\gamma_i^e)$

$$\Upsilon(\gamma_i^e) = Tr\{N_i^e \otimes P_{ii}(t) + M_i^e \otimes [AP^e(t-1)A^{\mathrm{T}} + Q] - \sum_{ii}^{e}(t-1)\} \,. \quad (2.55)$$

$\Upsilon(\gamma_i^e)$ with different reception probabilities is shown in Fig. 2.4. Thus, the condition (C3) is satisfied with low reception probabilities. This implies that the trace of the fusion EEC is unbounded. However, $\Upsilon(\gamma_i^e)(i = 1, 2)$ go to 0 with exponential rates when the eavesdropper has high successful decoding possibility, and the inequality (2.36) is not satisfied. Hence, the judgment criterion (C3) is effective.

2.5 Conclusion

This chapter has studied the privacy preservation problem of the secure dimensionality reduction fusion estimation. An AN injecting approach, which involves random information generated with the physical process and all system parameters, was proposed to achieve perfect expected secrecy. Some sufficient conditions, which depend on the transmitting probabilities and the successful decoding probabilities, were derived to diverge from the trace of the eavesdropper's EEC. Besides, the artificial noise power was designed to ensure the sufficient privacy. The simulation result demonstrated the effectiveness of the proposed methods.

References

1. Cheng TM, Malyavej V, Savkin AV (2010) Decentralized robust set-value state estimation in networked multiple sensor systems. Comput Math Appl 59(8):2636–2646
2. Fang J, Li H (2008) Joint dimension assignment and compression for distributed multisensor estimation. IEEE Signal Proc Lett 15:174–177
3. Leong AS, Quevedo DE, Dolz D, Dey S. (2017) Remote state estimation over packet dropping links in the presence of an eavesdropper. In arXiv [cs.SY]
4. Wu J, Yuan Y, Zhang H, Shi L (2013) How can online schedules improve communication and estimation tradeoff? IEEE Trans Sig Proc 61(7):1625–1631
5. Jazwioski A (1970) Stochastic processes and filtering theory. Academic, New York
6. Deng Z, Gao Y, Mao L, Li Y, Hao G (2005) New approach to information fusion steady-state Kalman filtering. Automatica 41:1695–1707
7. Anderson BDO, Moore JB (1979) Optimal filtering. Prentice-Hall
8. Sun S, Deng Z (2004) Multi-sensor optimal information fusion Kalman filter. Automatica 40(6):1017–1023
9. Zhang H, Cheng P, Shi L, Chen J (2015) Optimal denial-of-service attack scheduling with energy constraint. IEEE Trans Autom Cont 60(11):3023–3028
10. Rappaport TS (1996) Wireless communications: principles and practice. IEEE Publications
11. Xue F, Xie L, Kuma P (2005) The transport capacity of wireless networks over fading channels. IEEE Trans Inf The 51(3):834–847
12. Simon M, Alouini M (2005) Digital communication over fading channels. John Wiley & Sons
13. Ramirez F (2001) On the performance of ultra-wide-band signals in Gaussian noise and dense multipath. IEEE Trans Veh Tech 50(1):244–249

Chapter 3
Distributed Privacy-Preserving Fusion Estimation Using Homomorphic Encryption

3.1 Introduction

As mentioned in Chap. 2, injecting complex noises into communication channels can achieve high privacy. However, in some scenarios, the energy of sensors may come from portable power source, which means that the total energy consumption is limited to a certain range. In such cases, the sensors may not have enough power for extra operations, and complex noise strategies that consume plenty of extra energy may not be suitable in these systems.

Instead of complex noises, some cryptographic or perturbation methods are recommended for guaranteeing the privacy in such energy-constrained systems. Therefore, in this chapter, a traditional cryptographic approach called homomorphic encryption is adopted to protect the privacy of estimation systems. Specifically, the local state estimates are encrypted with public keys at sensor nodes such that the eavesdropper cannot intercept real information. Then, the legitimate user applies a FC to calculate encrypted fusion estimates. In this case, the user can obtain the original optimal fusion estimate by resorting to the homomorphisms and private keys. Notice that such encryption does not require any other energy consumption, because it only modifies the values of transmitted data and the encoding length remains the same.

3.2 Problem Formulation

In this chapter, we consider the following linear state-space model:

$$x(t+1) = Ax(t) + Bw(t), \tag{3.1}$$

$$y_i(t) = C_i x(t) + v_i(t) \ \ (i = 1, 2, \ldots, L), \tag{3.2}$$

where $x(t) \in \Re^n$ denotes the system state, and $y_i(t) \in \Re^{m_i}$ denotes the measurement of sensor i. $A \in \Re^{n\times n}$, $B \in \Re^{n\times b}$ and $C_i \in \Re^{m_i\times n}$ are time-invariant matrices

B. Chen et al., *Secure Fusion Estimation and Control for Cyber-Physical Systems Under Malicious Attacks*,https://doi.org/10.1007/978-981-97-9396-9_3

with proper dimensions. $w(t) \in \mathfrak{R}^b$ and $v_i(t) \in \mathfrak{R}^{m_i}$ are the system noises, and the statistical properties are given in Assumption 3.1. Meanwhile, we assume the system above is controllable and observable.

Assumption 3.1 The noises $w(t)$ and $v_i(t)$ are independent white Gaussian noises (WGNs) that satisfy

$$\begin{aligned}&\mathbb{E}\{[w^{\mathrm{T}}(t_1)v_i^{\mathrm{T}}(t_1)]^{\mathrm{T}}[w^{\mathrm{T}}(t_2)\ v_j^{\mathrm{T}}(t_2)]\}\\&\quad=\delta(t_1,t_2)\mathrm{diag}\{Q_w,\delta(i,j)Q_{v_i}\},\end{aligned} \tag{3.3}$$

where $\delta(i, j)$ is the delta function such that $\delta(i, j) = 1$ if $i = j$; otherwise, $\delta(i, j) = 0$.

Here, the distributed fusion structure is adopted thanks to its robustness and fault-tolerance. The sensors will calculate local state estimates (LSEs) $\hat{x}_i(t)$ based on their measurements $\{y_i(1), \cdots, y_i(t)\}$. Then, the distributed fusion estimate (DFE) $\hat{x}_f(t)$ will be calculated based on the following weighted fusion criterion:

$$\hat{x}_f(t) = \sum_{i=1}^{L} W_i(t)\hat{x}_i(t), \tag{3.4}$$

where $W(t) \triangleq [W_1(t) \cdots W_L(t)]$ is the weighting matrix.

Notice that the eavesdropper in this chapter is a coalition that colludes with all the parties to infer the DFE in the legitimate user. To avoid such information leakage, the Paillier homomorphic encryption (PHE) approach is employed to encrypt the transmission. More concretely, the transmitted data will be encrypted by the public keys, while the data under PHE is expressed by $[[\cdot]]$. In this case, the FC serves as a computation unit, and it just computes the encrypted DFE $[[\hat{x}_f(t)]]$. After calculating and transmitting the encrypted DFEs, the legitimate user can correctly decrypt them with private key and obtain the final estimate $\hat{x}_u(t)$.

Consequently, this chapter is to propose a privacy-preserving estimator for NMFS against the potential eavesdroppers. Besides, the optimal fusion estimate should be calculated by resorting to all the local information. Finally, the privacy requires to be proved with some concepts of cryptography.

3.3 Main Result

3.3.1 Privacy Goals

Before introducing PHE, we give some definitions about cryptographic privacy [2, 3]. First, the eavesdropper is assumed to be a semi-honest model, which is given by the following definition.

Definition 3.1 (*S*emi-honest Model). A party is semi-honest if it does not deviate from the steps of the protocol, but may store the exchanged messages and its internal coin tosses, as well as processes the data received in order to learn more information.

The eavesdropping attack above is passive, and it will not modify the transmitted data. Hence, the legitimate user can successfully decrypt by resorting to the private key. Then, to counter such a eavesdropper, we introduce the following privacy target called computational indistinguishability. Here, $\{0, 1\}^*$ is a sequence of bits with indefinite length and $X = \{X_n\}_{n\in\mathbb{N}}$ is a sequence of random variables X_n that ranges over strings of bits with a length polynomial in n.

Definition 3.2 (*C*omputational Indistinguishability). The ensembles $X = \{X_n\}_{n\in\mathbb{N}}$ and $Y = \{Y_n\}_{n\in\mathbb{N}}$ are computationally indistinguishable, denoted as $X \overset{c}{\equiv} Y$, if for every positive probabilistic polynomial-time algorithm $D : \{0, 1\}^* \to \{0, 1\}$, every positive polynomial $p(\cdot)$ with all sufficiently large n's, one has

$$|\mathrm{Prob}\{D(X_n) = 1\} - \mathrm{Prob}\{D(Y_n) = 1\}| < \frac{1}{p(n)}. \tag{3.5}$$

Then, the execution view requires to be defined.

Definition 3.3 (*E*xecution View). Let $f(\bar{x}) = (f_1(\bar{x}), \cdots, f_L(\bar{x}))$ be a deterministic polynomial-time function and Π a multi-party protocol that computes $f(\bar{x})$ with the input $\bar{x} = (x_1, \cdots, x_L)$. The view of the i-th party during an execution of Π using $\bar{x}$ is defined as

$$V_i^{\Pi}(\bar{x}) = (x_i, \text{coins}, M_i), \tag{3.6}$$

where the coins are the outcome of the party's internal coin toss, and M_i is the set of messages it has received. Then, the view of eavesdropper $V_e^{\Pi}(\bar{x})$ during an execution of Π is defined as

$$V_e^{\Pi}(\bar{x}) = (V_1^{\Pi}(x), \cdots, V_L^{\Pi}(x)). \tag{3.7}$$

Finally, the following privacy target called the multi-party privacy requires to be achieved.

Definition 3.4 (*M*ulti-party Privacy with Respect to Semi-honest Behavior). Since the eavesdropper can acquire the information from sensors 1 to L, one has $f_e(\bar{x}) = (f_1(\bar{x}), \cdots, f_L(\bar{x}))$. We say that the multi-party protocol Π computes $f(x)$ privately if there exists a polynomial-time algorithm, denoted by simulator S, such that

$$S(\bar{x}, f_e(\bar{x})) \overset{c}{\equiv} V_e^{\Pi}(\bar{x}). \tag{3.8}$$

3.3.2 Homomorphic Encryption

In order to achieve multi-party privacy, we applied PHE [4] to guarantee the secure estimation. The homomorphic cryptosystem is a system that allows the calculation on the encrypted data, and the effectiveness is ensured with a pair of keys including public key $pk = (N, g)$ and private key $sk = (\lambda, \delta)$. Besides, we use $\mathbb{Z}_N$ to represent the set of integers modulo N and $\mathbb{Z}_N^*$ represents the set of invertible numbers modulo N. Then, the encryption-decryption procedures that related to PHE are given as follows.
(1) KeyGen(λ): Input parameter λ, the output a public key pk and a private key sk. N represents the modulus which is the product of two numbers p and q, while g is an element with order N in the multiplicative group $\mathbb{Z}_{N^2}^*$. $\lambda = \text{lcm}(p-1, q-1)$ and $\delta = ((g^\lambda \bmod N^2 - 1)/N^{-1}) \bmod N$.
(2) Enc(m, pk): Input the public key and a message m, and output a ciphertext $c \triangleq [[m]]$. The detailed encryption process is given as

$$c = g^m r^N \bmod N^2, \tag{3.9}$$

where $r \in \mathbb{Z}_N$ is a non-zero random number.
(3) Dec(c, sk): Input the private key and a ciphertext c, and output the message that was encrypted as m. The detailed decrypt process is given as

$$m = ((c^\lambda \bmod N^2 - 1)/N)\delta \bmod N \tag{3.10}$$

Note that PHE can provide two fundamental homomorphisms: addition and multiplication. They are respectively denoted by "$\oplus$" and "$\otimes$" in this chapter, and "$\otimes$" is omitted when the type of multiplication can be inferred. The details of these operations are given as follows.
(1) Certain operations can calculate the addition of two messages:

$$\text{Dec}(\text{Enc}(m_1, pk) \oplus \text{Enc}(m_2, pk), sk) = m_1 + m_2. \tag{3.11}$$

The result can be proved by resorting to the multiplication of encrypted values:

$$\begin{aligned}[[m_1]]\,[[m_2]] &= g^{m_1} r_1^N \bmod N^2 \cdot g^{m_2} r_2^N \bmod N^2 \\ &= g^{m_1+m_2}(r_1 r_2)^N \bmod N^2 \\ &= [[m_1 + m_2]].\end{aligned} \tag{3.12}$$

(2) Certain operations can calculate the multiplication of two messages:

$$\text{Dec}(c \otimes \text{Enc}(m_1, pk), sk) = cm_1. \tag{3.13}$$

The proof is given as follows:

$$[[m_1]]^c = g^{cm_1}(r^c)^N \bmod N^2 = [[cm_1]]. \tag{3.14}$$

Besides, the privacy guarantees of PHE are based on the decisional composite residuosity assumption [4].

3.3.3 Privacy-Preserving Fusion

As mentioned before, LSEs should be calculated at sensor sides in the distributed fusion estimation structure. Instead of $\{y_i(1), \cdots, y_i(t-1)\}$ utilized for Luenberger observer, all the valid measurements $\{y_i(1), \cdots, y_i(t)\}$ at time t are gathered for Kalman filter in this chapter, because it can achieve high estimation performance according to the projection theory.

Specifically, the standard Kalman filter [5] is adopted to compute the LSE at each sensor:

$$\hat{x}_i(t) = A\hat{x}_i(t-1) + K_i(t)(y_i(t) - C_i A\hat{x}_i(t-1)), \tag{3.15}$$

where the optimal Kalman gain $K_i(t) \in \Re^{n\times m_i}$ can be calculated by the following form:

$$\begin{cases} P_i(t) = (\mathbf{I}_n - K_i(t)C_i)P_i^z(t), \\ P_i^z(t) = AP_i(t-1)A^{\mathrm{T}} + BQ_wB^{\mathrm{T}}, \\ K_i(t) = P_i^z(t)C_i^{\mathrm{T}}\left(C_iP_i^z(t)C_i^{\mathrm{T}} + Q_{v_i}\right)^{-1}. \end{cases} \tag{3.16}$$

Meanwhile, the estimation error cross covariance between the i-th and j-th LSEs can be recursively calculated as

$$P_{ij}(t) = (\mathbf{I}_n - K_i(t)C_i)(AP_{ij}(t-1)A^{\mathrm{T}} + BQ_wB^{\mathrm{T}})(\mathbf{I}_n - K_j(t)G_j)^{\mathrm{T}}. \tag{3.17}$$

In reality, the Kalman filter converges to the steady state just in several steps, and then the parameters will become constant. For brevity, we assume that the Kalman filter has already achieved steady state before system running. In this case, the corresponding steady-state parameters are given as follows:

$$\begin{cases} K_i \triangleq \lim_{t\to\infty} K_i(t), \\ P_{ii} \triangleq \lim_{t\to\infty} P_i(t), \\ P_{ij} \triangleq \lim_{t\to\infty} P_{ij}(t). \end{cases} \tag{3.18}$$

Based on the design of steady-state estimator, the computation burden can be efficiently reduced, because the operations on the real-time recursion are eliminated. Therefore, it saves more time to guarantee the successful realization of the PHE scheme.

Moreover, to obtain the optimal estimation performance, the weighting matrix will be calculated in the linear minimum variance sense. Based on the design of steady-state Kalman filter, the corresponding steady-state weighting matrix can be calculated as

$$\begin{cases} W = (e^{\mathrm{T}} P^{-1} e)^{-1} e^{\mathrm{T}} P^{-1}, \\ e \triangleq [\mathbf{I}_n^{\mathrm{T}} \cdots \mathbf{I}_n^{\mathrm{T}}]^{\mathrm{T}} \in \Re^{nL \times n}, \\ P \triangleq (P_{ij}) \in \Re^{nL \times nL}. \end{cases} \quad (3.19)$$

To counter the potential eavesdroppers, LSEs are going to be encrypted by PHE approach at sensor sides. It means that the encrypted LSEs will be transmitted in the communication channels instead of the raw LSEs. In this case, the fusion operations on the encrypted LSEs will become

$$[[\hat{x}_f(t)]] = W_1[[\hat{x}_1(t)]] \oplus \cdots \oplus W_L[[\hat{x}_L(t)]]. \quad (3.20)$$

Finally, the legitimate user will decrypt the received packet (3.20) with private key. The privacy-preserving fusion protocol using PHE is given in Algorithm 3.1. Furthermore, we give a theorem to show the privacy guarantee of the proposed secure estimator.

Algorithm 3.1 The Privacy-Preserving Fusion Protocol

For given $\hat{x}_i(0)$, $i = 1, \cdots, L$

Sensors:

1: Use KeyGen(λ) to calculate and distribute keys;
2: Calculate LSEs by steady-state Kalman filters (3.15)-(3.18);
3: Calculate encrypted LSEs $[[\hat{x}_i(t)]]$ with pk.

FC:

1: Calculate weighting fusion matrices by (3.19);
2: Calculate encrypted DFE $[[\hat{x}_f(t)]]$ by (3.20).

User:

1: Decrypt $[[\hat{x}_f(t)]]$ using sk to get $\hat{x}_u(t)$.

Theorem 3.1 *Algorithm 3.1 protects the computational privacy against the eavesdroppers given in Definition 3.1, while the legitimate user obtains the optimal DFE.*

Proof Denote the variables with $\widetilde{(\cdot)}$ as those obtained by the simulator, which are different from the execution views but follow the same distribution. Here, we shall analyze the execution views of eavesdropper and legitimate user. Since DFEs are independent of time, we just discuss the views at each time slot, and the index t is omitted for brevity.

First, we show the execution views of eavesdropper. The execution view of all the sensors indexed from 1 to L, i.e., V_s^Π, is given by [1]:

$$\begin{aligned} V_s^\Pi &= \{V_1^\Pi, \cdots, V_L^\Pi\} \\ &= (y_1(t), G_1, Q_{v_1}, [[\hat{x}_1(t)]], \text{coins}_1, \cdots \\ &\quad y_L(t), G_L, Q_{v_L}, [[\hat{x}_L(t)]], \text{coins}_L, pk), \end{aligned} \tag{3.21}$$

where coins_i $(i = 1, \cdots, L)$ are the random numbers that are exploited for the encryption process and key generation. Similarly, the execution view of FC is expressed by

$$V_f^\Pi = ([[\hat{x}_i(t)]], \cdots, [[\hat{x}_L(t)]], [[\hat{x}_f(t)]]). \tag{3.22}$$

Since the eavesdropper can intercept the transmissions among the sensors, FC, and legitimate user, its execution view can be denoted as

$$V_e^\Pi = (V_s^\Pi, V_f^\Pi). \tag{3.23}$$

Then, the simulator S_e can be acquired by generating $\widetilde{[[T_i]]}$, $\widetilde{[[\hat{x}_i(t)]]}$ and $\widetilde{\text{coins}_i}$, that is

$$\begin{aligned} S_e = &(y_1(t), G_1, Q_{v_1}, \widetilde{[[\hat{x}_1(t)]]}, \widetilde{\text{coins}_1}, \cdots \\ &y_L(t), G_L, Q_{v_L}, \widetilde{[[\hat{x}_L(t)]]}, \widetilde{\text{coins}_L}, \\ &\widetilde{[[\hat{x}_f(t)]]}, pk). \end{aligned} \tag{3.24}$$

Here, $\widetilde{\text{coins}_i}$ $(i = 1, \cdots, L)$ are generated according to the same distribution of coins_i and are independent of other parameters, which are same for $\widetilde{[[\hat{x}_i(t)]]}$ and $\widetilde{[[\hat{x}_f(t)]]}$. We can derive that $V_e^\Pi \overset{c}{\equiv} S_e$. Thus, the multi-party privacy w.r.t. the semi-honest eavesdropper can be ensured.

On the other hand, the legitimate user can decrypt the received information by resorting to the private key sk. Based on the homomorphisms (3.11) and (3.13), the decrypted message (3.20) can be described by the follow form:

$$\hat{x}_u(t) = \sum_{i=1}^{L} W_i \hat{x}_i(t). \tag{3.25}$$

It can be directly derived that $\hat{x}_u(t) \equiv \hat{x}_f(t)$. It indicates that the encrypted data $[[\hat{x}_f(t)]]$ can be successfully decrypted to $\hat{x}_f(t)$, which means the estimation performance of legitimate user is optimal. This completes the proof.

Remark 3.1 In the private estimator in [1], the aggregator employs the covariance intersection (CI) fusion approach to fuse the local estimates of managers. However, the cross relations between sensors are neglected. Hence, the fusion criterion weighted by matrix is considered in this chapter, which obtains higher performance thanks to the combination of estimation error cross covariances. Moreover, the systems are assumed to reach steady state in the beginning. It preserves the time for encryption, because the recursive computation of estimator gain is eliminated.

3.4 Simulation

In this section, we take the target-tracking system as an example, where two sensors are deployed to observe it. Due to the emergent demand for protecting the privacy of vehicles in recent years, the proposed secure estimator is employed. The state vector is denoted by $x(t) = \text{col}\{s(t), \dot{s}(t)\}$, where $s(t)$ and $\dot{s}(t)$ respectively stand for the position and velocity of a moving target. Based on the dynamics of the tracking system, the parameters in (3.1) and (3.2) can be described by [6]:

$$\begin{aligned} A &= \begin{bmatrix} 1.0 & T_s \\ 0.0 & 1.0 \end{bmatrix}, \; B = \begin{bmatrix} \frac{T_s^2}{2} \\ T_s \end{bmatrix}, \\ C_1 &= \begin{bmatrix} 1.0 & 0.5 \end{bmatrix}, C_2 = \begin{bmatrix} 1.0 & 1.0 \end{bmatrix}, \end{aligned} \tag{3.26}$$

where T_s is the sampling period set as $1.0s$. Meanwhile, the covariances of the WGNs $w(t)$, $v_1(t)$ and $v_2(t)$ are given as follows:

$$Q = 0.1, \; R_1 = 0.02, \; R_2 = 0.01. \tag{3.27}$$

To show the effectiveness, the proposed method is compared with the fusion estimator in [1], which applies CI as the fusion criterion. The trajectories of real states and estimates are shown in Figs. 3.1 and 3.2. It intuitively shows that the proposed secure DFE can precisely track the target under PHE, including position and velocity components. Due to the existence of stochastic noises, the estimation performance is assessed by the root mean square errors (RMSEs) with the Monte Carlo method over averagely 100 runs. With this metric, the performance of various estimators is plotted in Fig. 3.3. We can see from this figure that the RMSE of the proposed DFE is lower than that of all the LSEs, which implies the effectiveness of the fusion process. Meanwhile, the RMSE of DFE with CI in [1] is higher than that of the proposed DFE. It further demonstrates the better performance of the proposed fusion criterion.

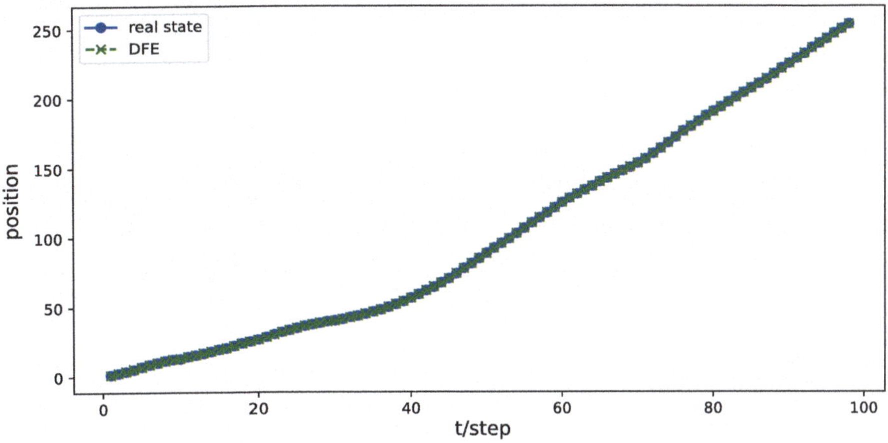

Fig. 3.1 The positions of real state and DFE

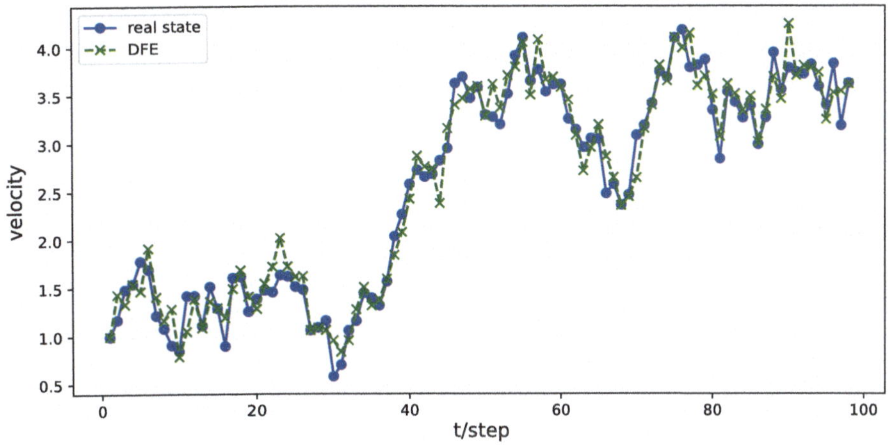

Fig. 3.2 The velocities of real state and DFE

3.5 Conclusion

In this chapter, the secure fusion estimation problem has been studied for a class of NMFS under eavesdropping attacks. By resorting to the PHE, we designed a distributed privacy-preserving fusion estimator, where the transmitted LSEs were encrypted with public keys at each sensor. Then, the encrypted LSEs were received and utilized for computing the encrypted DFEs in FC, which shows the invisibility of the actual DFEs. Then, based on the private key, the legitimate user successfully decrypts the data that was sent by FC, and it obtains the optimal DFEs. On the

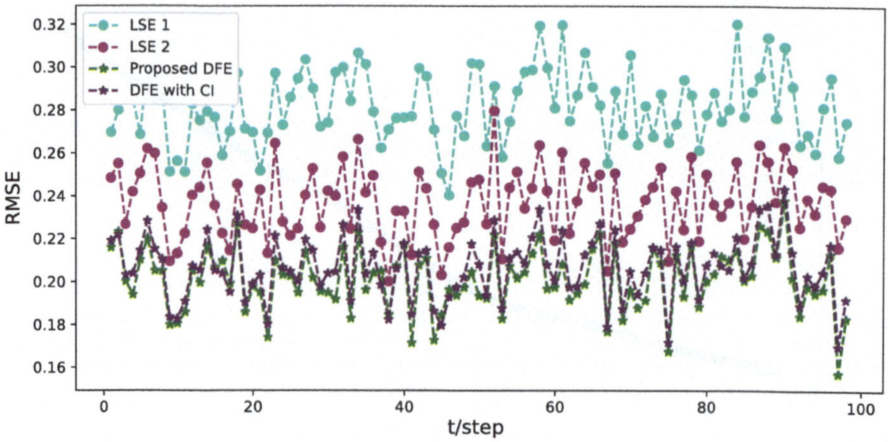

Fig. 3.3 The RMSEs of various estimators

other hand, the computational indistinguishability was guaranteed with some cryptographic definitions. Finally, a target-tracking system was employed to demonstrate the effectiveness of proposed method.

References

1. Emad S, Alanwar A, Alkabani Y, El-Kharashi MW, Sandberg H, Johansson KH (2022) Privacy guarantees for cloud-based state estimation using partially homomorphic encryption. In: 2022 European control conference (ECC)
2. Goldreich O (2003) Foundations of cryptography: Volume 1, basic tools. Cambridge University Press, Cambridge, UK
3. Goldreich O (2004) Foundations of cryptography: volume 2, basic applications. Cambridge University Press, Cambridge, UK
4. Paillier P (1999) Public-key cryptosystems based on composite degree residuosity classes. In: Proceedings of the 17th international conference on theory and application of cryptographic techniques, pp 223–238
5. Kalman RE (1960) A new approach to linear filtering and prediction problems. J Bas Eng 82(1):35–45
6. Yan X, Zhang Y, Xu D, Chen B (2022) Distributed confidentiality fusion estimation against eavesdroppers. IEEE Trans Aerospace Electron Syst 58(4):3633–3642

Chapter 4
FDI Attack Detection Using Fusion Estimation Methods

4.1 Introduction

In Chaps. 2 and 3, we introduced secure fusion estimation methods to counter eavesdropping attacks, ensuring the minimal impact on transmitted signals. While eavesdropping attacks are addressed, the focus is shifted to false data injection (FDI) attacks in this chapter. FDI attacks have the potential to easily manipulate system measurement data and control input commands, leading to the system performance degradation or irreparable harm. Hence, FDI attacks are considered to be the most common and dangerous. Timely detection and countermeasures are crucial in thwarting these attacks at an early stage, which is a pivotal aspect for ensuring the secure operation of the system.

Various methods for detecting FDI attacks have been developed to meet stringent safety and robustness requirements of CPSs [1, 2]. Specifically, the problem of detecting cyber-attacks on process control systems has been framed as a fault diagnosis challenge in [2]. Resilient attack detection estimators were formulated in [3] to identify false data. Combining graph theory with the smart grid system, an observer-based algorithm proposed in [4] leverages real-time measurements to detect and isolate FDI attacks. Notably, these FDI attack detection methods focus on identifying the presence of attack signals, but their response speed may fall short of practical system requirements, such as those for unmanned vehicles and critical infrastructures. Furthermore, the potential for artificially designed attacks to inflict greater damage underscores the significance of expediting the alarm response in attack detection–an area where few results have been reported.

Noticing that the design of FDI attack detectors based on estimation methods hinges on the performance of estimation and the accuracy of detection thresholds, this chapter explores the role of multi-sensor fusion estimation (MFE). MFE, which investigates the optimal utilization of information from multiple data sets to estimate parameters or processes [5], has emerged as a critical concern in information fusion.

B. Chen et al., *Secure Fusion Estimation and Control for Cyber-Physical Systems Under Malicious Attacks*,https://doi.org/10.1007/978-981-97-9396-9_4

Given its potential to enhance the estimation accuracy and reliability against attacks [6, 7], MFE offers a novel avenue for addressing the challenge of accelerating alarm response by refining the accuracy of detection thresholds.

This chapter delves into the alarm response issue of FDI attack detection using a multi-sensor fusion strategy, with a focus on attacks within the control input commands of CPSs. To facilitate real-time anomaly detection, a network of sensors strategically positioned send messages to a monitoring center (MC) via bandwidth-limited communication networks. Inspired by the fault detection filter design method in [8], a residual signal approximating the FDI attack signal at the actuator end is crafted. For quantized residual signals, the solution of the distributed fusion criterion is obtained by solving a convex optimization problem. Subsequently, fusion weights are designed using linear matrix inequality (LMI) methods to enhance the H_∞ performance of residual signals. The use of distributed fusion criteria yields a more accurate detection threshold. Comparative to a detection scheme relying on a single sensor, fusion detection significantly improves the system's warning speed. Finally, simulation examples are presented to validate the efficacy of the proposed method.

4.2 The Time-Invariant Case

4.2.1 Problem Statement

Consider a linear discrete-time system described by the following state-space model:

$$x(k+1) = Ax(k) + Bu(k) + E\omega(k), \tag{4.1}$$

or all $k \in N$, where $x(k) \in \mathfrak{R}^n$ represents the system state, $u(k) \in \mathfrak{R}^m$ is the control input, and $\omega(k) \in \mathfrak{R}^p$ denotes energy-bounded noises belonging to $l_2[0, \infty)$. A, B, E are constant matrices with appropriate dimensions. It is assumed that the pair (A, B) is controllable, and a feedback gain matrix $K \in \mathfrak{R}^{m\times n}$ has been designed such that $\Phi \triangleq A + BK$ is Hurwitz. In CPSs, the control signal sent to the actuator through communication networks may be susceptible to FDI attacks by a malicious attacker [9].

Specifically, the control signal received by the actuator can be expressed as:

$$\tilde{u}(k) = u(k) + a(k), \tag{4.2}$$

where $a(k) \in \mathfrak{R}^m$ represents an unknown attack signal. Clearly, if $a(k) \equiv 0$, then $\tilde{u}(k) = u(k)$, and the control signal is uncompromised. Alternatively, if $a(k)$ is nonzero, then the uncompromised control signal $u(k)$ is corrupted with malicious signal $a(k)$. Based on (4.1) and (4.2), the closed-loop system under FDI attacks is reconstructed as

$$x(k+1) = \Phi x(k) + Ba(k) + E\omega(k). \tag{4.3}$$

To detect the attack signal in (4.3) timely and accurately, it is proposed to deploy N sensors at different locations:

$$y_i(k) = C_i x(k) + D_i \omega(k), (i = 1, 2, \cdots, N), \tag{4.4}$$

where C_i and D_i are constant matrices with appropriate dimensions.

The local state estimator (LSE) based on the measurement $y_i(k)$ at the ith sensor is given by

$$\begin{cases} \hat{x}_i(k+1) = G_i \hat{x}_i(k) + M_i y_i(k), \\ z_i(k) = L_i \hat{x}_i(k), \end{cases} \tag{4.5}$$

where $G_i \in \mathfrak{R}^{n \times n}$, $M_i \in \mathfrak{R}^{n \times q_i}$, and $L_i \in \mathfrak{R}^{q \times n}$ are the estimator gains, while $z_i(k) \in \mathfrak{R}^q$ is the residual signal which is used to approach the attack signal $a(k)$. Define $e_i(k) = z_i(k) - a(k)$, $\eta_i(k) = \mathrm{col}\{x_i(k), \hat{x}_i(k)\}$, and $v(k) = \mathrm{col}\{\omega(k), a(k)\}$.

In this case, the augmented estimation error system is written as:

$$\begin{cases} \eta_i(k+1) = \bar{A}_i \eta_i(k) + \bar{B}_i v(k), \\ e_i(k) = \bar{C}_i \eta_i(k) + \bar{D}_i v(k), \end{cases} \tag{4.6}$$

where $\bar{A}_i = \begin{bmatrix} A & \mathbf{0} \\ M_i C_i & G_i \end{bmatrix}$, $\bar{B}_i = \begin{bmatrix} E & B \\ M_i D_i & \mathbf{0} \end{bmatrix}$, $\bar{C}_i = \begin{bmatrix} \mathbf{0} & L_i \end{bmatrix}$, and $\bar{D}_i = \begin{bmatrix} \mathbf{0} & -\mathbf{I} \end{bmatrix}$. For the LSE $\hat{x}_i(k)$, the key issue is to design the estimator gain matrices G_i, M_i, and L_i in (4.5) such that the estimation error system (4.6) ensures an H_∞ disturbance attenuation level bound (DALB).

When each residual signal $z_i(k)$ is transmitted to the MC via communication networks, $z_i(k)$ must be quantized before transmission. Due to the limited communication bandwidth, only finite-level quantized messages are sent to the MC. In this case, the logarithmic quantizer $\chi_i \triangleq q_i(\cdot)$ is introduced to quantize the residual signal $z_i(k)$, where the set of quantization level for the quantizer χ_i is represented by $U_i = \{\pm u_h^{(i)} : u_h^{(i)} = \rho_i^h u_0^{(i)}, h = 0, \pm 1, \pm 2, \cdots\} \cup \{0\} (0 < \rho_i < 1, u_0^{(i)} > 0)$, where ρ_i is the quantization density. Then, the logarithmic quantizer $q(\cdot)$ is defined by:

$$q_i(v) = \begin{cases} u_h^{(i)}, & \text{if } \frac{1}{1+\delta_i} u_h^{(i)} < v < \frac{1}{1-\delta_i} u_h^{(i)}, \\ 0, & \text{if } v = 0, \\ -q_i(-v), & \text{if } v < 0, \end{cases} \tag{4.7}$$

where $\delta_i = \frac{1-\rho_i}{1+\rho_i}$ $(0 < \delta_i < 1)$. It is known from [10] that $q_i(v)$ can be expressed as $q_i(v) = \left(1 + \tilde{\Delta}_i\right) v$ for certain $\tilde{\Delta}_i$ satisfying $\left|\tilde{\Delta}_i\right| \leq \delta_i$. Notice that a larger δ_i means a coarser quantizer χ_i, which implies that the length of the transmitted data packet will be decreased as the value δ_i increases. Therefore, it is reasonable to model the bandwidth constraints via δ_i as follows [11]:

$$\delta_i \geq \delta_i^0 \ (i = 1, 2, \cdots, N), \tag{4.8}$$

where δ_i^0 $(i = 1, 2, \cdots, N)$ is the minimum values satisfying the communication capacity of the channel i. Here, the quantization parameters δ_i $(i = 1, 2, \cdots, N)$ can be adjusted to satisfy the constraint condition (4.8).

Let $z_{r_i}(k)$ denote the local quantized estimate (LQE), then it follows from (4.7) that

$$z_{r_i}(k) = q_i(z_i(k)) = \left(1 + \tilde{\Delta}_i\right) z_i(k). \tag{4.9}$$

To minimize the error between the LQEs $z_{r_i}(k)$ and the FDI attack signal $a(k)$, the distributed H_∞ fusion of the LQEs is computed by:

$$z_{r_0}(k) = \sum_{i=1}^{N} W_i z_{r_i}(k), \tag{4.10}$$

where the weighted matrices $W_1, \ldots, W_N$ satisfying $\sum_{i=1}^{N} W_i = I$ will be constructed in Sect. 4.2.2. Furthermore, it follows from (4.10) that the fusion error $e(k) = z_{r_0}(k) - a(k)$ is given by

$$e(k) = \sum_{i=1}^{M} W_i \left(1 + \tilde{\Delta}_i\right) z_i(k) - a(k). \tag{4.11}$$

Define $\Upsilon(k) = \mathrm{col}\left\{\hat{x}_1(k), \cdots, \hat{x}_N(k), x(k)\right\}$. Then, one has by (4.3)-(4.5) and (4.11) that

$$\begin{cases} \Upsilon(k+1) = \tilde{A}\Upsilon(k) + \tilde{B}v(k), \\ e(k) = \tilde{C}\Upsilon(k) + \tilde{D}v(k), \end{cases} \tag{4.12}$$

where

$$\tilde{A} = \begin{bmatrix} G_1 & \mathbf{0} & \cdots & \mathbf{0} & M_1 C_1 \\ \mathbf{0} & G_2 & \cdots & \mathbf{0} & M_2 C_2 \\ \vdots & \vdots & \ddots & \vdots & \vdots \\ \mathbf{0} & \mathbf{0} & \cdots & G_N & M_N C_N \\ \mathbf{0} & \mathbf{0} & \cdots & \mathbf{0} & \Phi \end{bmatrix},$$

$$\tilde{B} = \begin{bmatrix} M_1 D_1 & \mathbf{0} \\ M_2 D_2 & \mathbf{0} \\ \vdots & \vdots \\ M_N D_N & \mathbf{0} \\ E & B \end{bmatrix}, \tilde{C} = \bar{C} + \sum_{i=1}^{N} \tilde{C}_{\tilde{\Delta}_i},$$

$$\tilde{C}_{\tilde{\Delta}_i} = \begin{bmatrix} \underbrace{\mathbf{0} \cdots \mathbf{0}}_{i-1} & W_i \tilde{\Delta}_i L_i & \underbrace{\mathbf{0} \cdots \mathbf{0}}_{N+1-i} \end{bmatrix},$$

$$\bar{C} = [\, W_1 L_1 \; W_2 L_2 \; \cdots \; W_N L_N \; \mathbf{0} \,], \; \tilde{D} = [\, \mathbf{0} \; -\mathbf{I} \,].$$

To detect the occurrence of FDI attacks, it is proposed to construct the following residual evaluation function for the fusion detector and each local detector:

$$\Theta_i(k) = \frac{1}{k} \left\{ \sum_{t=0}^{k} z_{r_i}^T(t) z_{r_i}(t) \right\}^{\frac{1}{2}}, \; i=0, \cdots, N. \tag{4.13}$$

Then, one chooses the residual evaluation threshold Θ_{th_i} as

$$\Theta_{th_i} = \sup_{v(k)\in l_2, a(k)=0} \mathbb{E}\{\Theta_i(k)\}, \; i=0, \cdots, N. \tag{4.14}$$

Based on (4.13) and (4.14), when the value of the residual evaluation function $\Theta_i(k)$ is larger than a given threshold Θ_{th_i}, an alarm of attacks will be triggered, i.e.,

$$\begin{aligned} &\Theta_i(k) > \Theta_{th_i} \Rightarrow \text{with FDI attacks} \Rightarrow \text{alarm}, \\ &\Theta_i(k) \le \Theta_{th_i} \Rightarrow \text{without FDI attacks}. \end{aligned} \tag{4.15}$$

Remark 4.1 Notice that $\Theta_0(k)$ in (4.13) and Θ_{th_0} in (4.14) represent the residual evaluation function and evaluation threshold of fusion detector, respectively. Moreover, the fusion detector and each local detector are realized by computing $\Theta_i(k)$ with (4.13) and Θ_{th_i} with (4.14), respectively, and then judging the condition (4.15).

Therefore, the primary objective is to design the LSE gains G_i, M_i and L_i in (4.5). The design process aims at minimizing the H_∞ DALB γ_i while adhering to constraint

$$\sup_{v(k)\neq 0} \frac{\mathbb{E}\left\{\|e_i(k)\|^2\right\}}{\|v(k)\|^2} < \gamma_i^2 \tag{4.16}$$

under the zero-initial condition, where $\gamma_i > 0$. Then, according to the quantization strategy (4.7) and the desired fusion estimation algorithm (4.10), the second objective is to find a set of optimal weighted matrices $W_1, \cdots, W_N$ such that, for the bandwidth constraint condition (4.8), the H_∞ DALB $\gamma > 0$ of the system (4.12) is minimal, i.e.,

$$\begin{aligned} &\{W_i, \cdots, W_N\} = \arg\min \gamma \\ &\text{s.t.} \left\|\tilde{C}(sI - \tilde{A})^{-1}\tilde{B} + \tilde{D}\right\|_\infty < \gamma \text{ and } (3.8). \end{aligned} \tag{4.17}$$

Remark 4.2 The local state estimators are fine-tuned to detect whether the FDI attacks are occurring. It should be noted that the threshold setting is crucial for the accuracy of these detection methods. Thus, the distributed sensor fusion in the form of (4.10) is introduced so that the responsiveness to FDI attacks can be improved by using fusion detectors compared with local detectors.

4.2.2 Main Result

In this section, optimal LSE gains G_i, M_i and L_i in (4.5) and optimal weighted matrices W_i $(i = 1, \cdots, N)$ in (4.10) will be presented in Theorem 4.1. Define:

$$\begin{cases} \Phi_{1_i} = \begin{bmatrix} \mathbf{0} & Y_i \end{bmatrix}, \Phi_{2_i} = \begin{bmatrix} \mathbf{0} & -\mathbf{I} \end{bmatrix}, \\ \Phi_{3_i} = \begin{bmatrix} P_{1_i} - R_i - R_i^{\mathrm{T}} & P_{2_i} - S_i - T_i \\ * & P_{3_i} - T_i - T_i^{\mathrm{T}} \end{bmatrix}, \\ \Phi_{4_i} = \begin{bmatrix} R_i^{\mathrm{T}}\Phi + X_i C_i & Z_i \\ S_i^{\mathrm{T}}\Phi + X_i C_i & Z_i \end{bmatrix}, \\ \Phi_{5i} = \begin{bmatrix} R_i^{\mathrm{T}}E + X_i D_i & R_i^{\mathrm{T}}B \\ S_i^{\mathrm{T}}E + X_i D_i & S_i^{\mathrm{T}}B \end{bmatrix}, \\ \Phi_{6i} = -\begin{bmatrix} P_{1_i} & P_{2_i} \\ * & P_{3_i} \end{bmatrix}. \end{cases} \tag{4.18}$$

Theorem 4.1 *For a given parameter $\delta_i > 0$, the optimal estimator gains G_i, M_i and L_i in (4.5) are computed as*

$$\begin{bmatrix} G_i & M_i \\ L_i & \mathbf{0} \end{bmatrix} = \begin{bmatrix} T_i^{-1} & \mathbf{0} \\ \mathbf{0} & \mathbf{I} \end{bmatrix} \begin{bmatrix} Z_i & X_i \\ Y_i & \mathbf{0} \end{bmatrix}, \tag{4.19}$$

where T_i, X_i, Y_i, and Z_i are obtained by solving the following convex optimization problem:

$$\min_{\mathrm{P}_{1_i}>0,\mathrm{P}_{3_i}>0,\mathrm{P}_{2_i},R_i,S_i,T_i,X_i,Y_i,Z_i} \hat{\gamma}_i$$
$$\text{s.t.} \begin{bmatrix} -\mathbf{I} & \mathbf{0} & \Phi_{1_i} & \Phi_{2_i} \\ * & \Phi_{3_i} & \Phi_{4_i} & \Phi_{5_i} \\ * & * & \Phi_{6_i} & \mathbf{0} \\ * & * & * & -\hat{\gamma}_i\mathbf{I} \end{bmatrix} < 0. \tag{4.20}$$

Then, by solving the following convex optimization problem, a set of optimal weighted matrices $W_i (i = 1, 2, \cdots, N)$ can be obtained:

$$\min_{\bar{P}>0} \hat{\gamma}$$
$$\text{s.t.} \begin{bmatrix} \Psi_1 & \Psi_2 & \mathbf{0} \\ * & \Psi_3 & \Psi_4 \\ * & * & \Psi_5 \end{bmatrix} < 0, \tag{4.21}$$

where

$$\Psi_1 = \mathrm{diag}\{-\varepsilon_N\mathbf{I}, \cdots, -\varepsilon_1\mathbf{I}\}, \ \Psi_2 = \begin{bmatrix} \mathbf{0} \ \Psi_6 \ \mathbf{0} \ \mathbf{0} \end{bmatrix},$$
$$\Psi_3 = \begin{bmatrix} -\tilde{P} & \tilde{P}\tilde{A} & \tilde{P}\tilde{B} & \mathbf{0} \\ * & -\tilde{P} & \mathbf{0} & \bar{C}^{\mathrm{T}} \\ * & * & -\hat{\gamma}\mathbf{I} & \tilde{D}^{\mathrm{T}} \\ * & * & * & -\mathbf{I} \end{bmatrix},$$
$$\Psi_4 = \begin{bmatrix} \mathbf{0} \\ \psi_7 \end{bmatrix}, \ \Psi_5 = \mathrm{diag}\{\varepsilon_1\mathbf{I}, \cdots, \varepsilon_N\mathbf{I}\},$$
$$\Psi_6 = \mathrm{col}\{\varepsilon_N H_N^{\mathrm{T}}, \cdots, \varepsilon_1 H_1^{\mathrm{T}}\}, \ \Psi_7 = [Q_1, \cdots, Q_N],$$
$$H_i = \mathrm{diag}\{\underbrace{\mathbf{0} \cdots \mathbf{0}}_{i-1} \ L_i^T\delta_i \ \underbrace{\mathbf{0} \cdots \mathbf{0}}_{N+1-i}\}, \ Q_i = [\underbrace{\mathbf{0} \cdots \mathbf{0}}_{i-1} \ W_i \ \underbrace{\mathbf{0} \cdots \mathbf{0}}_{N+1-i}].$$

Proof Construct the following Lyapunov candidate function: $V_i(k) = \eta_i^{\mathrm{T}}(k) \bar{P}_i\eta_i(k)$. Then, one has

$$\begin{aligned} \mathcal{V}_i(k) &= V_i(k+1) - V_i(k) \\ &\quad + \mathrm{E}\left\{e_i^{\mathrm{T}}(k)e_i(k)\right\} - \gamma_i^2\mathrm{E}\left\{v^{\mathrm{T}}(k)v(k)\right\}. \end{aligned} \tag{4.22}$$

Then, it follows from (4.6) and (4.22) that

$$\mathcal{V}_i(k) = \begin{bmatrix} \eta_i(k) \\ v(k) \end{bmatrix}^{\mathrm{T}} \begin{bmatrix} \Xi_{1_i} & \Xi_{2_i} \\ * & \Xi_{3_i} \end{bmatrix} \begin{bmatrix} \eta_i(k) \\ v(k) \end{bmatrix}, \tag{4.23}$$

where $\Xi_{1_i} = A_i^T\bar{P}_iA_i - \bar{P}_i + C_i^TC_i$, $\Xi_{2_i} = A_i^T\bar{P}_iB_i + C_i^TD_i$, and $\Xi_{3_i} = B_i^T\bar{P}_iB_i + D_i^TD_i - \gamma_i^2 I$. From the Schur complement lemma [12], $\mathcal{V}_i(k) < 0$ is equivalent to

$$\begin{bmatrix} -\mathbf{I} & \mathbf{0} & \bar{C}_i & \bar{D}_i \\ * & -\bar{P}_i & \bar{P}_i\bar{A}_i & \bar{P}_i\bar{B}_i \\ * & * & -\bar{P}_i & \mathbf{0} \\ * & * & * & -\gamma_i^2\mathbf{I} \end{bmatrix} < 0. \tag{4.24}$$

Then, it can be concluded from [13] that the inequlity (4.24) will be true if there exists a Π_i such that

$$\begin{bmatrix} -\mathbf{I} & \mathbf{0} & \bar{C}_i & \bar{D}_i \\ * & \psi_i & \Pi_i^{\mathrm{T}}\bar{A}_i & \Pi_i^{\mathrm{T}}\bar{B}_i \\ * & * & -\bar{P}_i & \mathbf{0} \\ * & * & * & -\gamma_i^2\mathbf{I} \end{bmatrix} < 0, \tag{4.25}$$

where $\psi_i = P_i - \Pi_i^{\mathrm{T}} - \Pi_i$.

Considering that $\bar{P}_i$ is positive definite, it can be further confirmed that Π_i is non-singular. We partition $\bar{P}_i$ and Π_i as follows:

$$\bar{P}_i = \begin{bmatrix} \bar{P}_{1_i} & \bar{P}_{2_i} \\ * & \bar{P}_{3_i} \end{bmatrix}, \Pi_i = \begin{bmatrix} \Pi_{1_i} & \Pi_{2_i} \\ \Pi_{3_i} & \Pi_{4_i} \end{bmatrix}. \tag{4.26}$$

Introduce the following matrices:

$$\Gamma_i = \begin{bmatrix} \mathbf{I} & \mathbf{0} \\ \mathbf{0} & \Pi_{4_i}^{-1}\Pi_{3_i} \end{bmatrix}, P_i = \begin{bmatrix} P_{1_i} & P_{2_i} \\ * & P_{3_i} \end{bmatrix} = \Gamma_i^T \bar{P}_i \Gamma_i. \tag{4.27}$$

Then, performing congruence transformation to (4.25) by diag $\{I, \Gamma_i, \Gamma_i, \mathbf{I}\}$, we can obtain

$$\begin{bmatrix} -\mathbf{I} & \mathbf{0} & \bar{\Phi}_{1_i} & \bar{D}_i \\ * & \bar{\Phi}_{2_i} & \bar{\Phi}_{3_i} & \bar{\Phi}_{4_i} \\ * & * & -P_i & \mathbf{0} \\ * & * & * & -\gamma_i^2\mathbf{I} \end{bmatrix} < 0, \tag{4.28}$$

where

$$\begin{aligned}
\bar{\Phi}_{1_i} &= \begin{bmatrix} \mathbf{0} & L_i\Pi_{4_i}^{-1}\Pi_{3_i} \end{bmatrix}, \bar{\Phi}_{2_i} = \begin{bmatrix} \bar{\Phi}_{5_i} & \bar{\Phi}_{6_i} \\ * & \bar{\Phi}_{7_i} \end{bmatrix}, \\
\bar{\Phi}_{3_i} &= \begin{bmatrix} \Pi_{1_i}^{\mathrm{T}}\Phi + \Pi_{3_i}^{T}M_iC_i & \Pi_{3_i}^{\mathrm{T}}G_i\Pi_{4_i}^{-1}\Pi_{3_i} \\ \Pi_{3_i}^{T}\Pi_{4_i}^{-\mathrm{T}}\Pi_{2_i}^{\mathrm{T}}\Phi + \Pi_{3_i}^{\mathrm{T}}M_iC_i & \Pi_{3_i}^{\mathrm{T}}G_i\Pi_{4_i}^{-1}\Pi_{3_i} \end{bmatrix}, \\
\bar{\Phi}_{4_i} &= \begin{bmatrix} \Pi_{1_i}^{\mathrm{T}}E + \Pi_{3_i}^{\mathrm{T}}M_iD_i & \Pi_{1_i}^{\mathrm{T}}B \\ \Pi_{3_i}^{\mathrm{T}}\Pi_{4_i}^{-\mathrm{T}}\Pi_{2_i}^{\mathrm{T}}E + \Pi_{3_i}^{\mathrm{T}}M_iD_i & \Pi_{3_i}^{\mathrm{T}}\Pi_{4_i}^{-\mathrm{T}}\Pi_{2_i}^{\mathrm{T}}B \end{bmatrix}, \\
\bar{\Phi}_{5_i} &= P_{1_i} - \Pi_{1_i} - \Pi_{1_i}^{\mathrm{T}}, \\
\bar{\Phi}_{6_i} &= P_{2_i} - \Pi_{2_i}\Pi_{4_i}^{-1}\Pi_{3_i} - \Pi_{3_i}^{\mathrm{T}}\Pi_{4_i}^{-1}\Pi_{3_i}, \\
\bar{\Phi}_{7_i} &= P_{3_i} - \Pi_{3_i}^{\mathrm{T}}\Pi_{4_i}^{-1}\Pi_{3_i} - \Pi_{3_i}^{\mathrm{T}}\Pi_{4_i}^{-T}\Pi_{3_i}.
\end{aligned}$$

Let

$$\begin{aligned} R_i &= \Pi_{1_i}, S_i = \Pi_{2_i}\Pi_{4_i}^{-1}\Pi_{3_i}, T_i = \Pi_{3_i}^{\mathrm{T}}\Pi_{4_i}^{-1}\Pi_{3_i}, \\ X_i &= \Pi_{3_i}^{\mathrm{T}} M_i, Y_i = L_i\Pi_{4_i}^{-1}\Pi_{3_i}, Z_i = \Pi_{3_i}^{\mathrm{T}} G_i \Pi_{4_i}^{-1}\Pi_{3_i}. \end{aligned} \tag{4.29}$$

Substituting (4.29) into (4.28), (4.20) can be obtained. Meanwhile, it is deduced from (4.27) and (4.29) that

$$\begin{aligned} \Pi_i &= \begin{bmatrix} R_i & S_i\Pi_{3_i}^{-1}\Pi_{4_i} \\ \Pi_{3_i} & \Pi_{4_i} \end{bmatrix}, G_i = \Pi_{3_i}^{-\mathrm{T}} Z_i \Pi_{3_i}^{-1}\Pi_{4_i}, \\ \bar{P}_i &= \Gamma_i^{-\mathrm{T}} P_i \Gamma_i^{-1}, M_i = \Pi_{3_i}^{-T} X_i, L_i = \Pi_{3_i}^{-1}\Pi_{4_i} Y_i. \end{aligned} \tag{4.30}$$

By substituting the parameters in (4.30) into the transfer function of the estimation error system (4.6) and considering the relationship $T_i = \Pi_{3_i}^T \Pi_{4_i}^{-1}\Pi_{3_i}$, we have

$$\begin{aligned} T_{yri}(s) &= L_i(sI - G_i)^{-1} M_i \\ &= Y_i\left(sI - T_i^{-1} Z_i\right)^{-1} T_i^{-1} X_i, \end{aligned} \tag{4.31}$$

which means that LSE parameters can be computed by (4.19).

Based on the LSE gains above, the procedure for solving convex optimization problem (4.21) will be given. It is concluded from the Discrete-time Bounded Real Lemma [14] that (4.12) satisfies H_∞ disturbance attenuation level γ if and only if there exits $\tilde{P} = \tilde{P}^{\mathrm{T}} > \mathbf{0}$ such that

$$\begin{cases} \tilde{A}^{\mathrm{T}}\tilde{P}\tilde{A} - \tilde{P} + \tilde{C}^{\mathrm{T}}\tilde{C} - \gamma^{-2}\Lambda_1\Lambda_2^{-1}\Lambda_1^{\mathrm{T}} < \mathbf{0}, \\ \Lambda_2 < \mathbf{0}, \end{cases} \tag{4.32}$$

where

$$\Lambda_1 = \left(\tilde{A}^{\mathrm{T}}\tilde{P}\tilde{B} + \tilde{C}^{\mathrm{T}}\tilde{D}\right), \Lambda_2 = \gamma^{-2}\left(\tilde{D}^{\mathrm{T}}\tilde{D} + \tilde{B}^{\mathrm{T}}\tilde{P}\tilde{B}\right) - \mathbf{I}.$$

By using the Schur complement lemma, (4.32) is equivalent to

$$\begin{bmatrix} -\tilde{P} & \tilde{P}\tilde{A} & \tilde{P}\tilde{B} & \mathbf{0} \\ * & -\tilde{P} & \mathbf{0} & \tilde{C}^{\mathrm{T}} \\ * & * & -\gamma^2\mathbf{I} & \tilde{D}^{\mathrm{T}} \\ * & * & * & -\mathbf{I} \end{bmatrix} < \mathbf{0}. \tag{4.33}$$

Let $F_{\tilde{\Delta}_i} \triangleq \tilde{\Delta}_i/\delta_i$, which satisfies $F_{\tilde{\Delta}_i} F_{\tilde{\Delta}_i}^{\mathrm{T}} \leq I$. Then, one has

$$\begin{bmatrix} \mathbf{0}\,\mathbf{0}\,\mathbf{0} & \mathbf{0} \\ \mathbf{0}\,\mathbf{0}\,\mathbf{0} & \tilde{C}^{\mathrm{T}}_{\tilde{\Delta}_i} \\ \mathbf{0}\,\mathbf{0}\,\mathbf{0} & \mathbf{0} \\ \mathbf{0}\,\mathbf{0}\,\mathbf{0} & \mathbf{0} \end{bmatrix} = \begin{bmatrix} \mathbf{0} \\ H_i \\ \mathbf{0} \\ \mathbf{0} \end{bmatrix} F_{\tilde{\Delta}} \left[\mathbf{0}\,\mathbf{0}\,\mathbf{0}\; Q_i^{\mathrm{T}} \right], \tag{4.34}$$

where $F_{\tilde{\Delta}} \triangleq \mathrm{diag}\left\{F_{\tilde{\Delta}_1}, F_{\tilde{\Delta}_2}, \cdots, F_{\tilde{\Delta}_N}, \mathbf{I}\right\}$.

Then, according to Lemma 4 of [15], the inequality (4.34) holds if and only if there exits a scalar $\varepsilon_i > 0$ such that

$$\begin{bmatrix} -\tilde{P} & \tilde{P}\tilde{A} & \tilde{P}\tilde{B} & \mathbf{0} \\ * & -\tilde{P} & \mathbf{0} & \bar{C}^{\mathrm{T}} \\ * & * & -\hat{\gamma}\mathbf{I} & \tilde{D}^{\mathrm{T}} \\ * & * & * & -\mathbf{I} \end{bmatrix} + \sum_{i=1}^{N} \left(\varepsilon_i \mathcal{H}_i \mathcal{H}_i^{\mathrm{T}} + \varepsilon_i^{-1} \mathcal{Q}_i \mathcal{Q}_i^{\mathrm{T}}\right) < \mathbf{0}, \tag{4.35}$$

where $\mathcal{H}_i = \mathrm{col}\,\{\mathbf{0}, H_i, \mathbf{0}, \mathbf{0}\}$,$\mathcal{Q}_i = \mathrm{col}\,\{\mathbf{0}, \mathbf{0}, \mathbf{0}, Q_i\}$, and $\hat{\gamma} = \gamma^2$.

By applying the Schur complement, premultiplying and postmultiplying by $\mathrm{diag}\,\{\varepsilon_N \mathbf{I}, \cdots, \varepsilon_1 \mathbf{I}, \mathbf{I}\}$ and its transpose, it is found that inequality (4.35) is equivalent to (4.21). This completes the proof. □

Remark 4.3 The optimization problems (4.20) and (4.21) in Theorem 4.1 are established in terms of LMI, and thus they can be directly solved by the function "mincx" in MATLAB LMI Toolbox [12].

Based on Theorem 4.1, the detection procedures for FDI attacks of linear system (4.3) under bandwidth constraint (4.8) are summarized in Algorithm 4.1.

Algorithm 4.1 Fusion-based FDI Attack Detection

1: Determine optimal LSE gains G_i, M_i and L_i by solving the optimization problem (4.20);
2: Determine the optimal weighted matrices W_i by solving the optimization problem (4.21);
3: Calculate residual threshold of MC Θ_{th_0} by (4.14);
Recursions for $k = 1, 2, 3...$
4: **for** $i := 1$ **to** N **do**
5: Calculate the LQE $z_{r_i}(k)$ by substituting (4.4) into (4.9);
6: **end for**
7: Calculate the MFE $z_{r_0}(k)$ by (4.10);
8: Calculate residual function of MC $\Theta_0(k)$ by (4.13);
9: **if** $\Theta_0(k) > \Theta_{th_0}$ **then**
10: An alert of attacks will be triggered;
11: **end if**
12: Return to Step 4.

4.2.3 Simulation

An F-404 aircraft engine system model with a sampling time of $T_0 = 0.1$s in [16] is used to demonstrate the effectiveness of proposed scheme, where three sensors are deployed to monitor the work status of the engine, and the residual signals are transmitted to the MC through communication networks.

Let the state be $x(k) = \text{col}\{x_1(k), x_2(k), x_3(k)\}$, where $x_1(k)$ and $x_2(k)$ denote the horizontal position, and $x_3(k)$ is the altitude of the aircraft. Then, the system matrices Φ, B and E in (4.3) are given by:

$$\Phi = \begin{bmatrix} 0.8673 & 0.0000 & 0.2022 \\ 0.0293 & 0.9763 & -0.0301 \\ 0.0259 & 0.0000 & 0.8032 \end{bmatrix},$$

$$B = \begin{bmatrix} 0.1 & 0 \\ 0.5 & 0 \\ -0.02 & 0 \end{bmatrix}, \quad E = \begin{bmatrix} 0.164 \\ 0.084 \\ 0.438 \end{bmatrix}$$

and the system noise $\omega(k)$ is taken as $\omega(k) = 0.05 \times \sin(2k) \times v(k) \times e^{-k/200}$, where $v(k)$ is uniformly distributed over $[-1, 1]$. The FDI attack signal $a(k)$ is set to be

$$a(k) = \begin{cases} 0.5 \times \cos(2k), & k \geq 15\text{s}, \\ 0, & \text{otherwise.} \end{cases}$$

Meanwhile, the measurement matrices in (4.4) are taken as

$$C_1 = \begin{bmatrix} 1 & 0 & 0 \\ 0 & 0 & 1 \end{bmatrix}, \quad D_1 = \begin{bmatrix} 0.2 & 0.1 \\ 0 & 0.4 \end{bmatrix},$$
$$C_2 = \begin{bmatrix} 1 & 1 & 0 \\ 0 & 1 & 1 \end{bmatrix}, \quad D_2 = \begin{bmatrix} 0.2 & 0.1 \\ 0 & 0.2 \end{bmatrix},$$
$$C_3 = \begin{bmatrix} 1 & 0 & 0 \\ 0 & 1 & 1 \end{bmatrix}, \quad D_3 = \begin{bmatrix} 0.2 & 0.1 \\ 0 & 0.2 \end{bmatrix}.$$

Then, the optimization problem (4.20) can be solved by using LMI Toolbox in MATLAB, and the corresponding local optimal estimator gains G_i, M_i and L_i in (4.5) are obtained.

The parameters of bandwidth constraints (4.8) are set as $\delta_1 = \delta_2 = \delta_3 = 0.5$. With the given parameters, the optimal weighted matrices are given by $W_1 = 0.3707$, $W_2 = 0.3116$, and $W_3 = 0.3176$ by solving the optimization problem (4.21). By using 400 times Monte Carlo tests without FDI attack, the thresholds for each LSE and MC are shown in Table 4.1, which shows that the fusion-based detection threshold is lower than the local detection threshold.

Table 4.1 Detection of FDI attacks

	Threshold	Detection time (s)
Local detector 1	2.5580×10^{-7}	3.7
Local detector 2	1.7713×10^{-7}	2.1
Local detector 3	1.5634×10^{-7}	1.7
Fusion detector	1.4846×10^{-7}	1.1

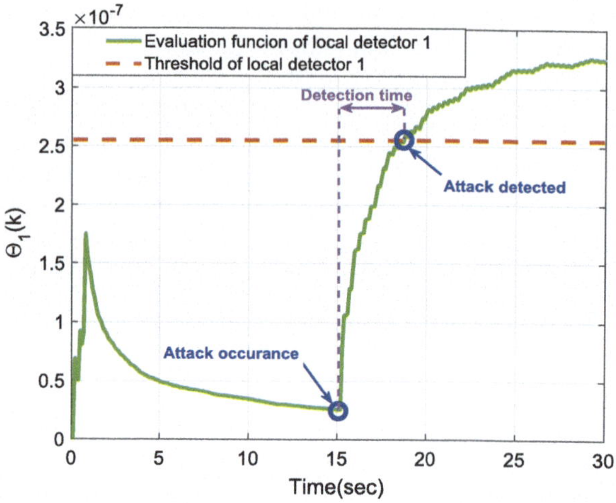

Fig. 4.1 Detection of local detector 1

Furthermore, by implementing Algorithm 4.2, the results of FDI attack detection are demonstrated in Figs. 4.1, 4.2, 4.3, 4.4, where the dotted line in Figs. 2.1, 2.2, 2.3 and 2.4 means the threshold, while the solid line represents each residual evaluation function. Meanwhile, based on different detectors, the time needed by detecting the attack after its occurrence is presented in Table 4.1. It can be seen from Figs. 4.1, 4.2, 4.3, 4.4 and Table 4.1 that fusion detector has a faster response speed than each local detector.

4.3 The Time-Varying Case

4.3.1 Problem Statement

Consider the following linear time-varying system:

$$x(k+1) = A(k)x(k) + B(k)u(k) + E(k)\omega(k), \tag{4.36}$$

where $x(k) \in \mathrm{R}^n$ is system state vector; $u(k) \in \mathrm{R}^z$ is system control input; $w(k) \in l_2[0, m]$ is the energy system subjected to bounded noise; $A(k)$, $B(k)$, and $E(k)$ are

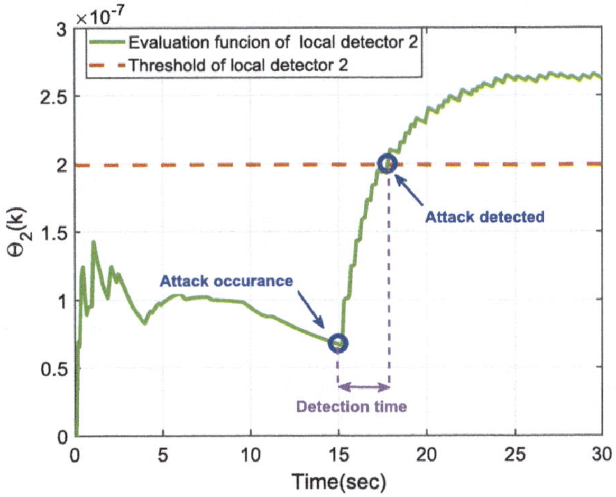

Fig. 4.2 Detection of local detector 2

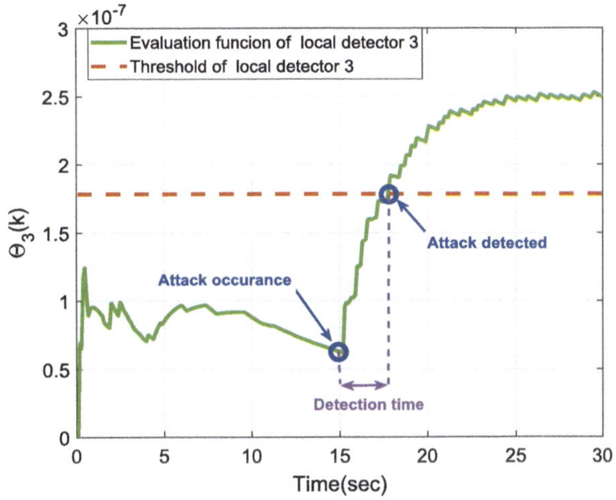

Fig. 4.3 Detection of local detector 3

time-varying matrices with appropriate dimensions. When the actuator is attacked by FDI, the actual control signal is

$$\tilde{u}(k) = u(k) + a(k),$$

where $a(k) \in l_2[0, m]$ represents energy bounded unknown FDI attack signal. The states of the system are observed in real time by deploying N sensors. The measurement output $y_i(k) \in R^m$ is modeled as:

$$y_i(k) = H_i(k)x(k) + D_{l_i}(k)\omega(k), i = 1, \cdots, N. \tag{4.37}$$

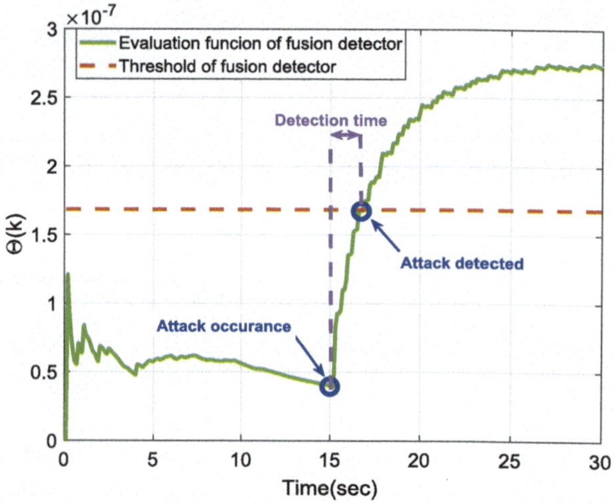

Fig. 4.4 Detection of fusion detector

The residual generator is designed as follows:

$$\begin{cases} r_i(k) = V_i(k)\chi_i(k), \\ \chi_i(k) = y_i(k+1) - H_i(k+1)A(k)\hat{x}_i(k) \\ \qquad - H_i(k+1)B(k)u(k), \\ \hat{x}_i(k+1) = A(k)\hat{x}_i(k) + B(k)u(k) \\ \qquad + L_i(k)\chi_i(k), \end{cases} \tag{4.38}$$

where the time-varying matrices $V_i(k)$ and $L_i(k)$ are gain matrices to be designed for residual generator, and r_i is residual signal. Define

$$\begin{cases} \xi(k) \triangleq \left[a(k),\ \omega^{\mathrm{T}}(k),\ \omega^{\mathrm{T}}(k+1)\right], \\ \tilde{r}_i(k) \triangleq r_i(k) - a(k), \\ e_i(k) \triangleq x(k) - \hat{x}_i(k). \end{cases}$$

Based on equations (4.36), (4.37) and (4.38), the following error system is established:

$$\begin{cases} e_i(k+1) = \bar{A}_i(k)e_i(k) + \bar{B}_i(k)\xi(k), \\ \tilde{r}_i(k) = \bar{C}_i(k)e_i(k) + \bar{D}_i(k)\xi(k), \end{cases} \tag{4.39}$$

where

$$\begin{cases} C_i(k) = H_i(k+1)A(k), \\ D_{a_i}(k) = H_i(k+1)B(k), \\ \bar{A}_i(k) = A(k) - L_i(k)C_i(k), \\ \bar{B}_i(k) = [B(k) - L_i(k)D_{a_i}(k), B_d(k) - L_i(k)D_{d_i}(k)], \\ \bar{C}_i(k) = V_i(k)C_i(k), \\ \bar{D}_i(k) = [V_i(k)D_{a_i}(k) - I, V_i(k)D_{d_i}(k)], \\ D_{d_i}(k) = [H_i(k+1)E(k),\ D_{l_i}(k+1)], \\ B_d(k) = [E(k),\ \mathbf{0}]. \end{cases}$$

The objective of this chapter is to design the gain matrices $V_i(k)$ and $L_i(k)$ to make the system (4.39) exponentially stable [8], and to give the anti-disturbance rejection ratio $\gamma_i > 0$ under the zero initial condition. Approximation error $\tilde{r}_i(k)$ between residual signal and FDI attack signal satisfies the following H_∞ performance indicators:

$$\sup_{\|\xi(k)\|_2 \neq 0} \mathrm{E}\left\{\frac{\|\tilde{r}_i(k)\|_2}{\|\xi(k)\|_2}\right\} < \gamma_i{}^2. \tag{4.40}$$

The residual signal $r_i(k)$ is quantified before passing through the communication network. Due to the limited communication bandwidth, the FC can only receive data information with limited quantification level. In this case, this chapter adopts the quantization strategy, and quantization sets with different quantization degrees [10] are expressed as:

$$\begin{aligned} U = &\left\{\pm u_p^{ij} : u_p^{ij} = \rho_{ij}{}^p u_0^{ij}, p = 0, \pm 1, \pm 2 \cdots\right\} \\ &\cup \{\pm u_0\} \cup \{0\}, 0 < \rho_{ij} < 1, u_0^{ij} > 0, \end{aligned} \tag{4.41}$$

where ρ_{ij} is the quantization degree at the j-th channel of residual signal $r_i(k)$. Define $\delta_{ij} = \frac{1-\rho_{ij}}{1+\rho_{ij}}, 0 < \delta_{ij} < 1$. It can be seen that with the increase of quantization degree (ρ_{ij} decreases), δ_{ij} and information loss of the original signal data become larger. Further, the quantizer [10] is expressed as:

$$q_{ij}(\nu) = \begin{cases} u_p^{ij}, & \frac{1}{1+\delta_{ij}} u_p^{ij} < \nu \leq \frac{1}{1-\delta_{ij}} u_p^{ij}, \\ 0, & \nu = 0, \\ -q_{ij}(-\nu), & \nu < 0. \end{cases} \tag{4.42}$$

Quantization error is defined by $\Delta_i(k) \triangleq \operatorname{diag}\{\Delta_{i1}(k), \cdots, \Delta_{iz}(k)\}$, $|\Delta_{ij}(k)| \leq \delta_{ij}$. Then the residual signal received by the fusion center can be expressed as

$$r_{q_i}(k) = (I + \Delta_i(k))r_i(k). \tag{4.43}$$

In order to improve the anti-disturbance ability of residual signal, fusion weight $W_i(k)$ is designed to meet conditions $\mathbf{0} \leq W_i(k) \leq \mathbf{I}$ and $\sum_{i=1}^{N} W_i(k) = \mathbf{I}$. The residual signal $r_{q_0}(k)$ after distributed fusion can be expressed as

$$r_{q_0}(k) = \sum_{i=1}^{N} W_i(k) r_{q_i}(k). \tag{4.44}$$

The augmented state error is defined as $\eta(k) \triangleq \text{col}\{e_1(k), \cdots, e_N(k)\}$ and the fusion error is defined as $\tilde{r}_0(k) \triangleq r_{q_0}(k) - a(k)$. Based on (4.39), (4.43) and (4.44), the following fusion error system is derived:

$$\begin{cases} \eta(k+1) = A_\eta(k)\eta(k) + B_\eta(k)\xi(k), \\ \tilde{r}_0(k) = \bar{C}_\eta(k)\eta(k) + \bar{D}_\eta(k)\xi(k), \end{cases} \tag{4.45}$$

where

$$\begin{cases} A_\eta(k) = \text{diag}\{\bar{A}_1(k), \cdots, \bar{A}_N(k)\}, \\ B_\eta(k) = \text{col}\{\bar{B}_1(k), \cdots, \bar{B}_N(k)\}, \\ \bar{C}_\eta(k) = C_\eta(k) + C_\delta(k),\ \bar{D}_\eta(k) = D_\eta(k) + D_\delta(k), \\ C_\eta(k) = [W_1(k)V_1(k)C_1(k) \cdots W_N(k)V_N(k)C_N(k)], \\ C_\delta(k) = [C_{\delta_1}(k) \cdots C_{\delta_N}(k)], \\ C_{\delta_i}(k) = W_i(k)\Delta_i(k)V_i(k)C_i(k), \\ D_\eta(k) = [D_\eta^s(k) \sum\limits_{i=1}^{N} W_i(k)V_i(k)D_{d_i}(k)], \\ D_\eta^s(k) = \sum\limits_{i=1}^{N} W_i(k)V_i(k)D_{a_i}(k) - \mathbf{I}, \\ D_\delta(k) = [D_\delta{}^e(k)\ D_\delta{}^\xi(k)], \\ D_\delta{}^e(k) = \sum\limits_{i=1}^{N} W_i(k)\Delta_i(k)V_i(k)D_{a_i}(k), \\ D_\delta{}^\xi(k) = \sum\limits_{i=1}^{N} W_i(k)\Delta_i(k)V_i(k)D_{d_i}(k). \end{cases}$$

After the fusion center obtains the residual signal $r_{q_i}(k)$, the following evaluation function $J_i(k)$ is adopted in the residual evaluation stage:

$$J_i(k) = \frac{1}{k}\left\{\sum_{i=0}^{N} r_{q_i}^{\mathrm{T}}(k) r_{q_i}(k)\right\}^{\frac{1}{2}}, i = 0, \cdots, N. \tag{4.46}$$

The corresponding detection threshold $J_{th,i}$ can be expressed as

$$J_{th,i} = \sup_{a(k)=0} J_i(k), i = 0, \cdots, N. \tag{4.47}$$

Since the occurrence of attack signals will make $J_i(k)$ increase, the decision logic to determine whether an attack occurs is as follows: $J_i(k) < J_{th,i}$, no attack detected; $J_i(k) \geq J_{th,i}$, trigger the attack warning. $i = 0$ indicates the fusion detection attack warning. The rest represents local attack warning against the i-th residual signal.

The problem to be solved in this chapter is as follows. Firstly, the gain matrices $V_i(k)$ and $L_i(k)$ are designed for CPSs whose actuator is attacked by FDI, and the quantized residual signal $r_{q_i}(k)$ is obtained. Secondly, the fusion error system (4.45) optimizes the fusion weight $W_i(k)$ under the given performance index of H_∞, so that the fusion detection method has a faster detection speed.

Remark 4.4 Improving the robustness of residual signals against external disturbances is the key to FDI attack detection in this chapter. Therefore, the purpose of fusion is to further improve the anti-disturbance ability of the fusion error system (4.45) under the given performance index γ. In the process of residual evaluation, the selection of threshold is an important task. If the detection threshold is too large, the detection speed will be reduced; if the detection threshold is too small, the false positive rate will be increased. Compared with the detection method of single sensor, the proposed fusion detection method can obtain a more accurate detection threshold under the same threshold selection criteria through the design of distributed fusion criteria, thus improving the warning speed.

4.3.2 Main Result

Before designing the gain matrices $V_i(k)$ and $L_i(k)$ of the residual generator in (4.38), the matrices are defined as following:

$$\begin{cases} \Gamma_i(k) = K_i(k) + \Omega_i(k)\Xi_i^{-1}(k)\Omega_i^{\mathrm{T}}(k), \\ \Omega_i(k) = C_i(k)Q_i(k)C_i^{\mathrm{T}}(k)\hat{V}_i^{\mathrm{T}}(k) + D_{a_i}(k)(-I + \hat{V}_i(k)D_{a_i}(k))^{\mathrm{T}} \\ \qquad + D_{d_i}(k)D_{d_i}^{\mathrm{T}}(k)\hat{V}_i^{\mathrm{T}}(k), \\ K_i(k) = C_i(k)Q_i(k)C_i^{\mathrm{T}}(k) + D_{d_i}(k)D_{d_i}^{\mathrm{T}}(k) + D_{a_i}(k)D_{a_i}^{\mathrm{T}}(k), \\ \Psi_i(k) = B_d(k)D_{d_i}^{\mathrm{T}}(k)\hat{V}_i^{\mathrm{T}}(k) + B(k)(\hat{V}_i(k)D_{a_i}(k) \\ \qquad - I)^{\mathrm{T}} + A(k)Q_i(k)C_i^{\mathrm{T}}(k)\hat{V}_i^{\mathrm{T}}(k), \\ Z_i(k) = C_i(k)Q_i(k)A^{\mathrm{T}}(k) + D_{a_i}(k)B^{\mathrm{T}}(k) + D_{d_i}(k)B_d^{\mathrm{T}}(k). \end{cases}$$

Lemma 4.1 *[8]: For error system (4.39), given constants $\beta_i > 0$, $\gamma_i > 0$ and the initial state-weighted matrix $Q_i(0) > 0$, there is a matrix $Q_i(k)$ such that:*

$$\begin{cases} Q_i(k+1) = \bar{A}_i(k)Q_i(k)\bar{A}_i^T(k) + \beta_i I + \bar{B}_i(k)\bar{B}_i^T(k) \\ \qquad + G_i(k)\Xi_i^{-1}(k)G_i^T(k), \\ \Xi_i(k) > 0, \end{cases} \tag{4.48}$$

where

$$\begin{cases} \Xi_i(k) = -\bar{C}_i(k)Q_i(k)\bar{C}_i^T(k) - \bar{D}_i(k)\bar{D}_i^T(k) + \gamma_i^2 I, \\ G_i(k) = \bar{A}_i(k)Q_i(k)\bar{C}_i^T(k) + \bar{B}_i(k)\bar{D}_i^T(k). \end{cases}$$

Then, the error system (4.39) is mean square exponentially stable. Under a given H_∞ performance index, the optimal gain matrices $\hat{V}_i(k)$ and $\hat{L}_i(k)$ can be expressed by

$$\begin{aligned} \hat{V}_i(k) = D_{a_i}^T(k)(C_i(k)Q_i(k)C_i^T(k) \\ + D_{a_i}(k)D_{a_i}^T(k) + D_{d_i}(k)D_{d_i}^T(k))^{-1}, \end{aligned} \tag{4.49}$$

$$\hat{L}_i(k) = (Z_i(k) + \Omega_i(k)\Xi_i^{-1}(k)\Psi_i^T(k))^T\Gamma_i^{-1}(k). \tag{4.50}$$

Proof Define the Lyapunov candidate function:

$$J_i(k) = e_i^{\mathrm{T}}(k)P_i(k)e_i(k), \tag{4.51}$$

where $P_i(k)$ is a symmetric positive definite matrix. When

$$\mathcal{J}_i(k) = J_i(k+1) - J_i(k) + e_i^{\mathrm{T}}(k)e_i(k) - \gamma_i^2\xi^{\mathrm{T}}(k)\xi(k) \tag{4.52}$$

is negative definite, the mean square index of the error system (4.39) is stable and meets the H_∞ performance index [1]. According to Schur's complement theorem [12], the stability condition is equivalent to

$$\bar{A}_i^{\mathrm{T}}(k)P_i(k+1)\bar{A}_i(k) + \bar{C}_i^{\mathrm{T}}(k)\bar{C}_i(k) - P_i(k) + \Lambda_{1_i}^{\mathrm{T}}(k)\Lambda_{2_i}^{-1}(k)\Lambda_{1_i}(k) < 0, \tag{4.53}$$

where $\Lambda_{2_i}(k) > 0$,

$$\begin{cases} \Lambda_{1_i}(k) = \bar{A}_i^{\mathrm{T}}(k)P_i(k+1)\bar{B}_i(k) + \bar{C}_i^{\mathrm{T}}(k)\bar{D}_i(k), \\ \Lambda_{2_i}(k) = \gamma_i^2 I - \bar{B}_i^{\mathrm{T}}(k)P_i(k+1)\bar{B}_i(k) - \bar{D}_i^{\mathrm{T}}(k)\bar{D}_i(k). \end{cases}$$

According to the proof of Theorem 2 in [8], when the system operates in the finite time domain $[0, m]$, the operator T is defined as the linear mapping from $(e_i(0), \xi(k))$ to $\tilde{r}_i(k)$, whose state space expression is (4.39). Then, there exists a adjoint operator T^* of T, which satisfies the following inner product relation:

$$\langle T(e_i(0), \xi(k)), \tilde{r}_i(k)\rangle = \left\langle (e_i(0), \xi(k)), T^*\tilde{r}_i(k)\right\rangle.$$

The state space equation of operator T^* is constructed by using the inner product relation above. For the adjoint system, according to (4.51)-(4.53), we can draw the following conclusion: if there is a positive definite matrix $Q_i(k) = P_i(m+1-k)$ making

$$\begin{aligned}&\bar{A}_i(k)Q_i(k)\bar{A}_i^{\mathrm{T}}(k) + \bar{B}_i(k)\bar{B}_i^{\mathrm{T}}(k)\\&+G_i(k)\Xi_i^{-1}(k)G_i^{\mathrm{T}}(k) - Q_i(k+1) < 0\end{aligned} \tag{4.54}$$

and $\Xi_i(k) > 0$ valid, the error system (4.39) is exponentially stable according to the norm equivalence condition of T and T^*. By introducing the constant β_i, inequality (4.54) is equivalent to Riccati equation (4.48). Detailed proof can be found in reference [8], and for the solution of $\hat{V}_i(k)$, we can expand $\Xi_i(k)$ to get

$$\begin{aligned}\Xi_i(k) = {\gamma_i}^2 I &- V_i(k)C_i(k)Q_i(k)C_i^{\mathrm{T}}(k){V_i}^{\mathrm{T}}(k)\\&-(V_i(k)D_{a_i}(k) - I)(V_i(k)D_{a_i}(k) - I)^{\mathrm{T}}\\&-V_i(k)D_{d_i}(k)D_{d_i}^{\mathrm{T}}(k){V_i}^{\mathrm{T}}(k).\end{aligned}$$

It can be seen that matrix $\Xi_i(k)$ is a quadratic function of $V_i(k)$, where $Q_i(k)$ is solved by (4.48), and is known at every time. Under the given H_∞ performance index, the optimization objective of improving the robustness of residual signal $r_i(k)$ is defined as: for any vector $\theta(k)$ with appropriate dimension, solve

$$\max\left\{\theta^{\mathrm{T}}(k)\Xi_i(k)\theta(k)\right\}. \tag{4.55}$$

Let $\partial(\theta^{\mathrm{T}}(k)\Xi_i(k)\theta(k))/\partial(V_i^{\mathrm{T}}(k)\theta(k)) = 0$, and then we can get

$$\begin{aligned}&-\theta^{\mathrm{T}}(k)V_i(k)(C_i(k)Q_i(k)C_i^{\mathrm{T}}(k)\\&\quad+D_{a_i}(k)D_{a_i}^{\mathrm{T}}(k) + D_{d_i}(k)D_{d_i}^{\mathrm{T}}(k))\\&\quad+\theta^{\mathrm{T}}(k)D_{a_i}^{\mathrm{T}}(k) = 0.\end{aligned} \tag{4.56}$$

When the rows of $C_i(k)$ are full rank, we can get

$$\begin{aligned}&\partial^2(\theta^{\mathrm{T}}(k)\Xi_i(k)\theta(k))/\partial(V_i^{\mathrm{T}}(k)\theta(k))^2 =\\&-(C_i(k)Q_i(k)C_i^{\mathrm{T}}(k) + D_{a_i}(k)D_{a_i}^{\mathrm{T}}(k)\\&+D_{d_i}(k)D_{d_i}^{\mathrm{T}}(k))^{\mathrm{T}} < 0.\end{aligned} \tag{4.57}$$

So (4.49) is the optimal solution of optimization problem (4.55). In order to solve the optimal $L_i(k)$, optimization problem $\min\left\{\theta^{\mathrm{T}}(k)Q_i(k+1)\theta(k)\right\}$ is formulated. Similarly, let $\partial(\theta^{\mathrm{T}}(k)Q_i(k+1)\theta(k))/\partial(L_i^{\mathrm{T}}(k)\theta(k)) = 0$. Then, derive the optimal gain matrix $\hat{L}_i(k)$ as (4.50). Therefore, the proof of Lemma 4.1 is completed.

Remark 4.5 The residual generator in this chapter is a special form of the fault detection filter (FDF) in reference [8]. Since the residual signal in literature [8] depends on the gain matrix corresponding to the fault signal in the measurement model and the measured data is not damaged by the FDI attack on the actuator, the FDF cannot generate the residual for detecting the FDI attack signal on the actuator.

Lemma 4.2 *[17]: Define matrices A, H, E and F with appropriate dimensions, where $FF^T < I$. X is a symmetric positive definite matrix, and α is any normal number which satisfies the condition $\alpha^{-1}I - EXE^T > 0$. Then, the following inequality holds.*

$$(A + HFE)X(A + HFE)^T \leq A(X^{-1} - \alpha E^T E)^{-1}A^T + \alpha^{-1}HH^T. \tag{4.58}$$

After solving the optimal gain matrix of the residual generator, the fusion criterion is further designed to make $r_{q_0}(k)$ have better suppression effect on $\xi(k)$. Before giving the main result, the matrices are defined as follows:

$$\begin{cases} W(k) = [W_1(k) \cdots W_N(k)], \\ \Theta(k) = \gamma^2 I + \Theta_V(k), \\ C_\Delta(k) = Q(k)(C_\Delta^V(k))^T(k)\hat{\Delta}, \\ C_\Delta^V(k) = diag\{V_1(k)C_1(k), \cdots, V_N(k)C_N(k)\}, \\ \hat{\Delta} = diag\{\varepsilon\delta_1, \cdots, \varepsilon\delta_N\}, \\ \delta_i = diag\{\delta_{i1}, \cdots, \delta_{iz}\}, \\ D_\Delta(k) = (D_\Delta^V(k))^T\hat{\Delta}, \\ D_\Delta^V(k) = col\left\{\bar{D}_1^V(k), \cdots, \bar{D}_N^V(k)\right\}, \\ \bar{D}_i^V(k) = \left[V_i(k)D_{a_i}(k), V_i(k)D_{d_i}(k)\right], \\ \Pi_1 = A_\eta(k)Q(k)C_\eta^T(k) + B_\eta(k)D_\eta^T(k), \\ \Pi_2 = A_\eta(k)Q(k)(C_\Delta(k))^T + B_\eta(k)(D_\Delta^V(k))^T, \\ \Theta_L(k) = \Pi_1(k)(-\Theta_V(k) - \alpha W(k)W^T(k))\Pi_1^T(k) + \alpha^{-1}\Pi_2(k)\Pi_2^T(k). \end{cases}$$

Theorem 4.2 *Define the optimal gain matrices $\hat{V}_i(k)$ and $\hat{L}_i(k)$ of the local residual generator, H_∞ performance index γ of the fusion error system (4.45) and initial matrix $Q(0)$. Then, the optimal weighted fusion weight $W_i(k)$ is determined by the following convex optimization problem:*

$$\begin{aligned} &\min_{0 \leq W_i(k) \leq \mathbf{I}, 0 < \varepsilon, \Theta_V(k) < 0} \quad Tr\ \{\Theta_V(k)\} \\ &s.t. \begin{bmatrix} -\varepsilon\mathbf{I} & W^T(k) & \mathbf{0} & \mathbf{0} & \mathbf{0} \\ * & -\Theta(k) & C_\eta(k)Q(k) & D_\eta(k) & \mathbf{0} \\ * & * & -Q(k) & \mathbf{0} & C_\Delta(k) \\ * & * & * & -\mathbf{I} & D_\Delta(k) \\ * & * & * & * & -\varepsilon\mathbf{I} \end{bmatrix} < 0, \\ &\sum_{i=1}^{N} W_i(k) = \mathbf{I}, \end{aligned} \tag{4.59}$$

where $Q(k)$ can be obtained by the following recursion:

$$\begin{aligned} Q(k) = &\beta\mathbf{I} + A_\eta(k-1)Q(k-1)A_\eta^T(k-1) \\ &+ B_\eta(k-1)B_\eta^T(k-1) + \Theta_L(k-1). \end{aligned} \tag{4.60}$$

Proof Based on Lemma 4.1, for a given H_∞ performance index γ, there exists a $Q(k) > 0$ such that

$$\begin{aligned} Q(k+1) > & A_\eta(k)Q(k)A_\eta^{\mathrm{T}}(k) \\ & + B_\eta(k)B_\eta^{\mathrm{T}}(k) + G_\eta(k)\Xi^{-1}(k)G_\eta^{\mathrm{T}}(k), \end{aligned} \tag{4.61}$$

where $\Xi(k) > 0$,

$$\begin{cases} G_\eta(k) = A_\eta(k)Q(k)\bar{C}_\eta^{\mathrm{T}}(k) + B_\eta(k)\bar{D}_\eta^{\mathrm{T}}(k), \\ \Xi(k) = -\bar{C}_\eta(k)Q(k)\bar{C}_\eta^{\mathrm{T}}(k) - \bar{D}_\eta(k)\bar{D}_\eta^{\mathrm{T}}(k) + \gamma^2\mathbf{I}. \end{cases}$$

The mean square index of the fusion error system (4.45) is stable and satisfies the H_∞ performance. Similar to the solution of the gain matrix $\hat{V}_i(k)$ and $\hat{L}_i(k)$, for the selection of fusion weight $W_i(k)$, formulate the optimization problem as

$$\max\left\{\phi^{\mathrm{T}}(t)\Xi(k)\phi(k)\right\}, \tag{4.62}$$

where $\phi(k)$ is any vector with appropriate dimension. Due to the existence of quantization error, it is difficult to obtain the expression of the optimal fusion weight $W_i(k)$ by solving the quadratic problem. In this case, introduce matrix $\Theta_V(k) < 0$, which makes the following conditions hold

$$\bar{C}_\eta(k)Q(k)\bar{C}_\eta^{\mathrm{T}}(k) + \bar{D}_\eta(k)\bar{D}_\eta^{\mathrm{T}}(k) < \gamma^2\mathbf{I} + \Theta_V(k). \tag{4.63}$$

Therefore, the optimization problem (4.62) is equivalent to $\min\left\{\phi^{\mathrm{T}}(k)\Theta(k)\phi(k)\right\}$, where $\Theta(k) = \gamma^2\mathbf{I} + \Theta_V(k) > 0$. For the positive definite matrix $\Theta(k)$, the following inequality exists:

$$\phi^{\mathrm{T}}(k)\Theta(k)\phi(k) \le \lambda_{max}\left\{\phi(k)\phi^{\mathrm{T}}(k)\right\}\operatorname{Tr}\left\{\Theta(k)\right\}. \tag{4.64}$$

Then, the optimization problem (4.62) can be converted into $\min \operatorname{Tr}\{\Theta(k)\}$. Since γ is a given value, then $\min \operatorname{Tr}\{\Theta(k)\}$ is equivalent to $\min \operatorname{Tr}\{\Theta_V(k)\}$. According to Schur's complement lemma [17], (4.63) is equivalent to

$$\begin{bmatrix} -\Theta(k) & \bar{C}_\eta(k)Q(k) & \bar{D}_\eta(k) \\ * & -Q(k) & \mathbf{0} \\ * & * & -\mathbf{I} \end{bmatrix} < 0. \tag{4.65}$$

Note that there are uncertainties introduced to quantization in $\bar{C}_\eta(k)$ and $\bar{D}_\eta(k)$, which gives $F_{\Delta_{ij}}(k) = \frac{\Delta_{ij}(k)}{\delta_{ij}}$ and $F_{\Delta_{ij}}(k)F_{\Delta_{ij}}^{\mathrm{T}}(k) \le \mathbf{I}$. Meanwhile,

$$\begin{bmatrix} \mathbf{0} & C_\delta(k)Q(k) & D_\delta(k) \\ \mathbf{0} & \mathbf{0} & \mathbf{0} \\ \mathbf{0} & \mathbf{0} & \mathbf{0} \end{bmatrix} = \begin{bmatrix} W(k) \\ \mathbf{0} \\ \mathbf{0} \end{bmatrix} F_\Delta(k)Z(k), \tag{4.66}$$

where

$$\begin{cases} Z(k) = \left[\mathbf{0}\ \Delta C_{\Delta}^{V}(k)Q(k)\ \Delta D_{\Delta}^{V}(k)\right], \\ \Delta = \text{diag}\left\{\delta_1, \cdots, \delta_N\right\}, \\ F_{\Delta}(k) = \text{diag}\left\{F_{\Delta_{11}}(k), \cdots, F_{\Delta_{N_z}}(k)\right\}. \end{cases}$$

According to S-procedure [18], let $\mathbf{W}(k) = \text{col}\{W(k), \mathbf{0}, \mathbf{0}\}$. Then, inequality (4.65) is true if and only if there is a scalar $\varepsilon > 0$ that makes the following inequality hold

$$\begin{bmatrix} -\Theta(k) & \bar{C}_{\eta}(k)Q(k) & \bar{D}_{\eta}(k) \\ * & -Q(k) & \mathbf{0} \\ * & * & -\mathbf{I} \end{bmatrix} + \varepsilon Z^{\mathrm{T}}(k)Z(k) + \varepsilon^{-1}\mathbf{W}(k)\mathbf{W}^{\mathrm{T}}(k) < 0. \tag{4.67}$$

The convex optimization problem (4.59) is obtained by using Schur's complement lemma and multiplying the left and the right by diag $\{\mathbf{I}, \mathbf{I}, \mathbf{I}, \mathbf{I}, \varepsilon\mathbf{I}\}$ to eliminate the nonlinear terms in the LMI.

In addition, before solving convex optimization problem (4.59), a positive definite matrix $Q(k)$ satisfying condition (4.61) should be given. Since $\Delta_{ij}(k)$ is unknown, compared with the design process of local residual generator, unknown items need to be processed further after the normal $\beta > 0$ is introduced.

$$G_{\eta}(k)\Xi^{-1}(k)G_{\eta}^{\mathrm{T}}(k). \tag{4.68}$$

Expanding $G_{\eta}(k)$ gives

$$\begin{aligned} G_{\eta}(k) &= A_{\eta}(k)Q(k)C_{\eta}^{\mathrm{T}}(k) + B_{\eta}(k)D_{\eta}^{\mathrm{T}}(k) \\ &\quad + A_{\eta}(k)Q(k)(C_{\Delta}(k))^{\mathrm{T}}F_{\Delta}(k)W^{\mathrm{T}}(k) \\ &\quad + B_{\eta}(k)(D_{\Delta}^{V}(k))^{\mathrm{T}}F_{\Delta}(k)W^{\mathrm{T}}(k) \\ &= \Pi_1(k) + \Pi_2(k)F_{\Delta}(k)W^{\mathrm{T}}(k). \end{aligned} \tag{4.69}$$

According to the inequality (4.63), $-\Theta_V(k) < \Xi(k)$ can be obtained. Then based on lemma 4.2, we can conclude that given a normal number α satisfying condition $\alpha^{-1}I + W^{\mathrm{T}}(k)\Theta_V^{-1}(k)W(k)$, we have

$$\begin{aligned} &G_{\eta}(k)\Xi^{-1}(k)G_{\eta}^{T}(k) < -G_{\eta}(k)\Theta_V^{-1}(k)G_{\eta}^{T}(k) \\ &\leq \Theta_L(k) = \alpha^{-1}\Pi_2(k)\Pi_2^{\mathrm{T}}(k) + \Pi_1(k)(-\Theta_V(k) - \alpha W(k)W^{\mathrm{T}}(k))\Pi_1^{\mathrm{T}}(k), \end{aligned} \tag{4.70}$$

where $\Theta_V(k)$ is obtained by solving the convex optimization problem (4.59). Based on (4.61), (4.70) and normal number β, $Q(k)$ can be obtained from Riccati equation (4.60). Thus, Theorem 4.2 is proved.

Remark 4.6 Ignoring the quantization of residual signal will inevitably lead to the degradation of fusion performance. In order to improve the robustness of residual signals after fusion, this chapter transforms the selection problem of fusion weight $W_i(k)$ into a convex optimization problem in LMI form by introducing an unknown matrix $\Theta_V(k)$, and solves the problem by using the"$mincx$" function in Matlab LMI toolbox. Note that the optimization problem min Tr $\{\Theta_V(k)\}$ is similar to minimizing the H_∞ performance index in a linear time-invariant system [1], so the performance improvement of the residual signal H_∞ depends on whether a more negative-definite $\Theta_V(k)$ can be obtained, even if Tr$\{\Theta_V(k)\}$ is smaller.

Based on Lemma 4.1 and Theorem 4.2, the fusion attack detection algorithm for linear time-varying systems is implemented in Algorithm 4.2.

Algorithm 4.2 FDI Attack Signal Detection Based on Distributed Fusion

1: Select appropriate β, γ and the matrix $Q(0)$;
2: Calculate the gain matrices $\hat{V}_i(k)$ and $\hat{L}_i(k)$ by (4.49) and (4.50), respectively;
3: Calculate the optimal weight matrix $W_i(k)$ by (4.59);
4: Obtain the fusion detection threshold based on 50 Monte Carlo experiments with (4.46) and (4.47);
5: Calculate the local residual signal $r_i(k)$ by (4.38) and calculate the residual signal $r_{q0}(k)$ (4.44) at each moment;
6: Calculate the evaluation function of $r_{q0}(k)$ by (4.46);
7: **if** $J_0(k) \geq J_{th0}$ **then**
8: Attack signals are detected and an early warning is issued;
9: **else**
10: Jump to Step2 to continue running;
11: **end if**

4.3.3 Simulation

Considering a moving target system monitored by multiple sensors, the target motion process is represented by position and velocity. Define system state vector $x(k) \triangleq \text{col}\left\{S_x(k), V_x(k), S_y(k), V_y(k)\right\}$, where $S_x(k)$ and $S_y(k)$ are the position coordinates of the target on the X and Y axes, respectively, and $V_x(k)$ and $V_y(k)$ are the corresponding velocities, respectively. When the control signal is tampered with, the target movement trajectory is represented by the following state-space model [19]:

$$x(k+1) = \begin{bmatrix} 1 & f(k) & 0 & 0 \\ 0 & 1 & 0 & 0 \\ 0 & 0 & 1 & f(k) \\ 0 & 0 & 0 & 1 \end{bmatrix} x(k) + \begin{bmatrix} 0 \\ f(k) \\ 0 \\ f(k) \end{bmatrix} [u(k) + a(k)] + \begin{bmatrix} 0.4 & 0.2 & 0.2 & 0.1 \end{bmatrix}^{\mathrm{T}} \omega(k),$$

where the sampling period is set to $f(k) = 0.9 + 0.1\sin(k)$, and the feedback control signal is $u(k) = \begin{bmatrix} 0 & 0.01 & 0 & -0.01 \end{bmatrix} x(k)$. The noise $\omega(k) = 0.25\mu(k)\sin(1.5\mu(k))$ that affects the system during time interval [0, 200] can be generated by *rand* function of MATLAB. The movement of the target is monitored in real time by three sensors, and the corresponding measurement matrices are as follows:

$$H_1(k) = \begin{bmatrix} 0.5 & 0.7 & 0 & 0 \\ 0 & 0 & 0.9 & 0.6 \end{bmatrix}, \; D_{l_1}(k) = \begin{bmatrix} 0.2 \\ 0.3 \end{bmatrix},$$

$$H_2(k) = \begin{bmatrix} 0.9 & 0.8 & 0 & 0 \\ 0 & 0 & 0.5 & 0.1 \end{bmatrix}, \; D_{l_2}(k) = \begin{bmatrix} 0.2 \\ 0.4 \end{bmatrix},$$

$$H_3(k) = \begin{bmatrix} 0 & 0.2 & 0 & 0.2 \\ 0.8 & 0 & 0.8 & 0 \end{bmatrix}, \; D_{l_3}(k) = \begin{bmatrix} 0.3 \\ 0.3 \end{bmatrix}.$$

For fusion detection, let $\gamma_i = \gamma = 0.8$, $\beta_i = \beta = 0.01$, and initial matrices $Q_i(0) = 0.1\mathbf{I}$ and $Q(0) = 0.1\mathbf{I}$, where the subscript i represents the initial value setting of the local residual generator.

Let $\delta_{11} = \delta_{21} = \delta_{31} = 0.2$, and define unknown FDI attack signal $a(k) = 0.04 \sin(k/15)$ for the time during [0, 200]. Set the quantization degree of the three quantization channels as $\delta_{ij} = 0.3$, and set the noise suffered by the system as $\omega(k) = 0.75\mu(k)\sin(1.5\mu(k))$ when the threshold is selected to highlight the advantages of fusion detection. The FDI attack signal suffered by the system is set as

$$a(k) = \begin{cases} 0.05, & 100 < k < 200, \\ 0, & otherwise. \end{cases} \tag{4.71}$$

Algorithm 4.2 was executed, and the detection results are shown in Fig. 4.5, where the solid line represents the evaluation function curve corresponding to the residual signal $r_{q_i}(k)$, and the dashed line represents the detection threshold. The specific detection time can be seen in Table 4.2. It can be seen that fusion detection has a faster warning speed than the detection method using local residual signal. It is worth noting that the degree of quantification affects the change of fusion weight. When the quantization degree is the same as $\delta_{ij} = 0.3$, the corresponding weights can be seen in Table 4.2. When the quantization degree is $\delta_{11} = \delta_{21} = 0.2$ and $\delta_{31} = 0.3$, the corresponding weights are 0.4601, 0.4276 and 0.1123, respectively. In other words, the fusion criteria designed in this chapter can effectively assign greater weights to channels with less data loss and further improve the reliability of the detection system.

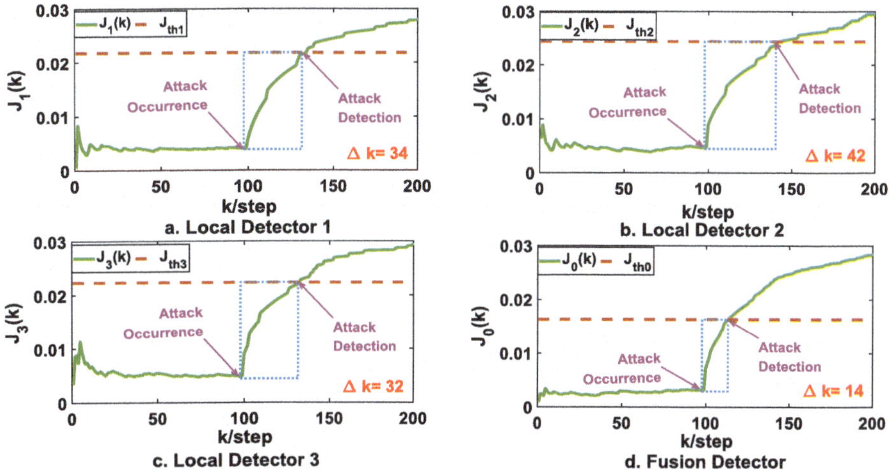

Fig. 4.5 FDI attack detection

Table 4.2 Attack Signal Detection

Detector	δ	Mean weight value	Threshold value	Detection time (s)
Local detector 1	0.3	0.3026	0.0219	34
Local detector 2	0.3	0.3305	0.0245	42
Local detector 3	0.3	0.3669	0.0224	32
Fusion detector	*	*	0.0164	14

4.4 Conclusion

This chapter studies the FDI attack detection of CPSs with limited bandwidth. Firstly, based on the given H_∞ performance index, an explicit expression of the unknown gain matrix of the local residual generator was given, and the residual signal which can approximate the attack signal is obtained. Furthermore, the detection method based on distributed fusion was adopted to design the fusion weight to improve the anti-disturbance ability of the residual signal, which greatly improves the warning speed of the detection system in the residual evaluation stage. Finally, simulation examples were given to verify the effectiveness of the proposed algorithm.

References

1. He X, Wang Z, Ji YD, Zhou DH (2011) Robust fault detection for networked systems with distributed sensors. IEEE Trans Aeros Electron Syst 47(1):166–177
2. Hwang I, Kim S, Kim Y, Seah CE (2010) A survey of fault detection, isolation, and reconfiguration methods. IEEE Trans Cont Syst Tech 18(3):636–653

3. Guan Y, Ge X (2018) Distributed attack detection and secure estimation of networked cyber-physical systems against false data injection attacks and jamming attacks. IEEE Trans Sig Inf Proc Over Netw 4(1):48–59
4. Luo X, Yao Q, Wang X, Guan X (2018) Observer-based cyber attack detection and isolation in smart grids. Int J Electr Power Energy Syst 101:127–138
5. Li XR, Zhu Y, Wang J, Han C. (2003) Optimal linear estimation fusion. I. Unified fusion rules. IEEE Trans Inf Theo 49(9):2192–2208
6. Chen B, Ho DWC, Hu G, Yu L (2018) Secure fusion estimation for bandwidth constrained cyber-physical systems under replay attacks. IEEE Trans Cybernet 48(6):1862–1876
7. Chen B, Ho DWC, Zhang WA, Yu L. (2019) Distributed dimensionality reduction fusion estimation for cyber-physical systems under DoS attacks. IEEE Trans Syst Man, and Cybernet Syst 49(2):455–468
8. Li Y, Zhong M (2010) On designing robust H_∞ fault detection filter for linear discrete time-varying systems with multiple packet dropouts (in chinese). Acta Automatic Sinica 36(12):1788–1796
9. Jin X, Haddad WM, Yucelen T (2017) An adaptive control architecture for mitigating sensor and actuator attacks in cyber-physical systems. IEEE Trans Automat Cont 62(11):6058–6064
10. Fu M, Xie L (2005) The sector bound approach to quantized feedback control. IEEE Trans Automat Cont 50(11):1698–1711
11. Chen B, Yu L, Zhang WA, Wang H (2014) Distributed H_∞ fusion filtering with communication bandwidth constraints. Signal Proc 96:284–289
12. Boyd S, El Ghaoui L, Feron E, Balakrishnan V (1994) Linear matrix inequalities in system and control theory. Soc Ind Appl Math
13. de Oliveira MC, Bernussou J, Geromel JC (1999) A new discrete-time robust stability condition. Syst and Cont Lett 37(4):261–265
14. de Souza CE, Xie L (1992) On the discrete-time bounded real lemma with application in the characterization of static state feedback H_∞ controllers. Syst Cont Lett 18(1):61–71
15. Yang F, Wang Z, Ho DW, Liu X (2006) Robust H_2 filtering for a class of systems with stochastic nonlinearities. IEEE Trans Circ Syst II: Express Briefs 53(3):235–239
16. Chen B, Hu G, Zhang WA, Yu L (2016) Distributed mixed H_2/H_∞ fusion estimation with limited communication capacity. IEEE Trans Autom Cont 61(3):805–810
17. Xie L, Soh YC, de Souza CE (1994) Robust Kalman filtering for uncertain discrete-time systems. IEEE Trans Automatic Cont 39(6):1310–1314
18. Bu X, Dong H, Wang Z, Liu H (2019) Non-fragile distributed fault estimation for a class of nonlinear time-varying systems over sensor networks: the finite-horizon case. IEEE Trans Sig Inf Proc Over Netw 5(1):61–69
19. Chen B, Ho DWC, Zhang WA, Yu L (2017) Networked fusion estimation with bounded noises. IEEE Trans Autom Cont 62(10):5415–5421

Chapter 5
Secure H_∞ Fusion Estimation Against FDI Attacks Based on Intermediate Variable

5.1 Introduction

Chapter 2 focuses on how to protect data privacy during transmission to prevent eavesdropping, while Chaps. 3 and 4 investigate rapid detection methods for malicious attacks. However, when a system is attacked and the attack is undetected, it is crucial to accurately estimate the true state of the system using the compromised data. In recent years, methods for secure state estimation have been categorized into three types based on the location of FDI attacks: sensor attacks, actuator attacks, and joint actuator-sensor attacks. Focusing on the scenarios where both actuators and sensors are compromised, Lu et al. [1] introduced a novel Luenberger-like observer to treat attacks as the components of an augmented system. Similarly, our prior work [2] addressed the estimation and state reconstruction in joint actuator-sensor attacks through a convex optimization approach. In the context of CPSs, both transmitted measurements and control messages in communication networks face increasing vulnerability to malicious attacks, underscoring the need for robust secure estimation algorithms. This chapter is motivated by the challenge of designing these algorithms specifically for actuator and sensor attacks. Moreover, compared with centralized approaches, distributed fusion offers enhanced robustness, flexibility, and reliability. These advantages have spurred extensive research into distributed fusion estimation over recent years [3]. Nevertheless, developing secure fusion estimation (SFE) for CPSs under multi-pass FDI attacks, such as simultaneous actuator and sensor attacks, continues to be a significant challenge.

Motivated by the analysis above, this chapter primarily aims to investigate the secure H_∞ fusion estimation problem for linear discrete-time CPSs subjected to simultaneous actuator and sensor attacks. In such scenarios, both control messages and local sensor measurements are vulnerable to the compromise caused by malicious entities. Based on the methodologies from our previous work [2], we introduce local intermediate estimators to reconstruct states and counteract FDI attacks. We achieve the estimation gains through a convex optimization framework, readily solvable using standard software. Additionally, we develop a secure H_∞ fusion structure

B. Chen et al., *Secure Fusion Estimation and Control for Cyber-Physical Systems Under Malicious Attacks*, https://doi.org/10.1007/978-981-97-9396-9_5

that integrates optimal weighting rules, aiming to improve the accuracy of both system state and actuator attack estimations. Finally, the effectiveness of our proposed methods is illustrated through a practical example involving a maneuvering vehicle.

5.2 Problem Formulation

Consider a discrete-time CPS where the sensors are distributed throughout the physical system, and each sensor only collects partial state information:

$$\begin{cases} x(k+1) = Ax(k) + B(u(k) + a_u(k)), \\ y_i(k) = C_i x(k) + D_i a_{yi}(k), \end{cases} \tag{5.1}$$

where $i = 1, \ldots, N$. The system state at time k is denoted by $x(k) \in \Re^n$, and the control input is represented by $u(k) \in \Re^m$. The term $a_u(k) \in \Re^m$ signifies the attacks targeting the actuator at time k. The output of the system from the ith sensor cluster at time k is $y_i(k) \in \Re^{p_i}$, while $a_{yi}(k) \in \Re^{q_i}$ represents the attacks directed at the ith sensors. The matrices A, B, C_i, and D_i are with appropriate dimensions.

Assumption 5.1 The variations in the malicious attacks, $a_u(k)$ and $a_{yi}(k)$, are energy-bounded. Specifically, the changes in the attack vectors, defined as $\Delta a_u(k) \triangleq a_u(k) - a_u(k-1)$ and $\Delta a_{yi}(k) \triangleq a_{yi}(k) - a_{yi}(k-1)$, belong to the $l_2[0, \infty)$ space. This implies that the energy of the deviations in these attack vectors over time is finite.

The local estimation (LE) of states and malicious attacks for (5.1) can be defined as $\hat{z}_i^+(k) \triangleq \text{col}\{\hat{x}_i(k), \hat{a}_{ui}(k), \hat{a}_{yi}(k)\}$ and the following estimator is given:

$$\begin{cases} \hat{z}_i^+(k+1) = G_1 \hat{z}_i^+(k) + G_2 u(k) \\ \qquad\qquad + L_i(y_i(k+1) - \hat{y}_i(k+1)), \\ \hat{y}_i(k+1) = G_3 \hat{z}_i^+(k+1), \end{cases} \tag{5.2}$$

where parameter matrices G_1, G_2, G_3 and gain L_i are to be determined in the next section.

For system state and actuator attack, we give the following definitions:

$$\begin{cases} z^-(k) \triangleq \text{col}\{x(k), a_u(k)\}, \\ \hat{z}_i^-(k) \triangleq \text{col}\{\hat{x}_i(k), \hat{a}_{ui}(k)\}, \end{cases} \tag{5.3}$$

and the secure fusion estimate of $z^-(k)$:

$$\hat{z}^-(k) = \sum_{i=1}^{L} W_i \hat{z}_i^-(k), \tag{5.4}$$

where the fusion weighting matrix W_i is to be determined. Consequently, the problem to be addressed is summarized as follows:

Problem 5.1 Design optimal gains L_i in (5.2) and weighting fusion matrices W_i in (5.4) such that the estimation error of each LE and SHFE of $\hat{z}^-(k)$ are asymptotically stable with the minimum H_∞ attack variation attenuation level bounds.

Remark 5.1 The proposed distributed secure fusion estimation problem encompasses both actuator and sensor attacks. Notably, there are no additional constraints (such as boundedness, duration, expectation, or variance) imposed on the attack signals $a_u(k)$ and $a_{yi}(k)$, other than the stipulation that their deviations are energy-bounded. Practically, this assumption is justified since subtle and less noticeable attacks are often preferred by adversaries for enhanced concealment.

5.3 Main Result

In this section, we focus on designing local intermediate estimators tailored for the system described in (5.1). The primary aim is to accurately estimate both the system states and the characteristics of the attacks. Additionally, we propose an optimal fusion algorithm. This algorithm is specifically developed to enhance the performance of secure estimation, ensuring more reliable and robust estimations in the presence of potential attacks.

Firstly, an intermediate variable is introduced to design the local estimator:

$$\tau(k) = a_u(k-1) - \mu B^{\mathrm{T}} x(k), \tag{5.5}$$

where μ is an adjustable parameter. Then, based on (5.1) and (5.5), the dynamics of $\tau(k)$ can be derived as

$$\begin{aligned} \tau(k+1) =&(\mathbf{I} - \mu B^{\mathrm{T}} B)\tau(k) - \mu B^{\mathrm{T}} B u(k) \\ &+ \mu B^{\mathrm{T}}(\mathbf{I} - A - \mu B B^{\mathrm{T}})x(k) - \mu B^{\mathrm{T}} B \Delta a_u(k). \end{aligned} \tag{5.6}$$

We define $\xi(k) \triangleq \mathrm{col}\{x(k), \tau(k)\}$ for the purpose of fusion estimation for states and actuator attacks, and the equivalent model of (5.1) and (5.6) is obtained:

$$\begin{cases} \xi(k+1) = \tilde{A}\xi(k) + \tilde{B}u(k) + M\Delta a_u(k), \\ y_i(k) = \tilde{C}_i\xi(k) + D_i a_{yi}(k), \end{cases} \tag{5.7}$$

where

$$\begin{cases} \tilde{A} = \begin{bmatrix} A + \mu B B^{\mathrm{T}} & B \\ \mu B^{\mathrm{T}}(\mathbf{I} - A - \mu B B^{\mathrm{T}}) & \mathbf{I} - \mu B^{\mathrm{T}} B \end{bmatrix}, \\ \tilde{B} = \begin{bmatrix} B \\ -\mu B^{\mathrm{T}} B \end{bmatrix}, M = \begin{bmatrix} \mathbf{I} - \mu B^{\mathrm{T}} B \\ \mathbf{I} \end{bmatrix}, \tilde{C}_i = \begin{bmatrix} C_i & \mathbf{0} \end{bmatrix}. \end{cases}$$

Subsequently, for each local sensor attack, it follows from (5.1) and (5.7) that

$$\begin{cases} z_i(k+1) = \bar{A}_i z_i(k) + \bar{B}_i u(k) + \bar{M}_i \Delta a_i(k+1), \\ y_i(k) = \bar{C}_i(k) z_i(k), \end{cases} \tag{5.8}$$

where

$$\begin{cases} z_i(k) = \text{col}\{\xi(k), a_{yi}(k)\}, \\ \Delta a_i(k) = \text{col}\{\Delta a_u(k-1), \Delta a_{yi}(k)\}, \\ \bar{A}_i = \begin{bmatrix} \tilde{A} & \mathbf{0} \\ \mathbf{0} & \mathbf{I}_{qi} \end{bmatrix}, \bar{B}_i = \begin{bmatrix} \tilde{B} \\ \mathbf{0} \end{bmatrix}, \\ \bar{M}_i = \begin{bmatrix} M & \mathbf{0} \\ \mathbf{0} & \mathbf{I}_{qi} \end{bmatrix}, \bar{C}_i = \begin{bmatrix} \tilde{C}_i & D_i \end{bmatrix}. \end{cases}$$

Consequently, the local intermediate estimator for the ith sensor is proposed as follows:

$$\begin{cases} \hat{z}_i(k+1) = \bar{A}_i \hat{z}_i(k) + \bar{B}_i u(k) + L_i(y_i(k+1) - \hat{y}_i(k+1)), \\ \hat{y}_i(k+1) = \bar{C}_i \bar{A}_i \hat{z}_i(k) + \bar{C}_i \bar{B}_i u(k), \end{cases} \tag{5.9}$$

where $\hat{z}_i(k) = \text{col}\{\hat{\xi}_i(k), \hat{a}_{yi}(k)\}$, $\hat{\xi}_i(k) = \text{col}\{\hat{x}_i(k), \hat{\tau}_i(k)\}$, and $L_i = \text{col}\{L_{1i}, L_{2i}\}$ are local estimation gains to be determined.

Theorem 5.1 Suppose that Assumption 5.1 holds. For given μ, the estimation errors of LEs are asymptotically stable with the minimum attack variation attenuation level bound if there exist a symmetrical matrix $P_i > 0$ and a matrix H_i such that the following convex optimization problem has feasible solutions:

$$\begin{aligned} &\min_{P_i>0, \gamma_i^\star>0, H_i} \quad \gamma_i^\star \\ &s.t. \begin{bmatrix} -P_i & P_i\bar{A}_i - H_i\bar{C}_i\bar{A}_i & P_i\bar{M}_i - H_i\bar{C}_i\bar{M}_i \\ * & \mathbf{I} - P_i & \mathbf{0} \\ * & * & -\gamma_i^\star \mathbf{I} \end{bmatrix} < 0. \end{aligned} \tag{5.10}$$

Then the local intermediate estimator gains L_i are calculated by $L_i = P_i^{-1} H_i$. Furthermore, a group of optimal weighting matrices $W_1, \ldots, W_N$ for local estimation (5.9) can be obtained if there exist symmetrical matrix $P_0 > 0$ and matrix H_0 such that the following optimization problem has feasible solutions:

$$\begin{aligned} &\min_{P_0>0, \gamma^\star>0, H_0} \quad \gamma^\star \\ &s.t. \begin{bmatrix} -P_0 & H_0 A_F & H_0 M_F \\ * & \mathbf{I} - P_0 & \mathbf{0} \\ * & * & -\gamma^\star \mathbf{I} \end{bmatrix} < 0, \end{aligned} \tag{5.11}$$

where

$$\begin{cases} A_F = \text{diag}\{\tilde{A}_{L_1}, \ldots, \tilde{A}_{L_N}\}, \\ M_F = \text{diag}\{M_{L_1}, \ldots, M_{L_N}\}, \\ \tilde{A}_{L_i} = \tilde{A} - L_{1i}\tilde{C}_i\tilde{A}, \\ M_{L_i} = \left[-L_{1i}D_i \; M - L_{1i}\tilde{C}_i M \; -L_{1i}D_i\right], \end{cases}$$

and $W \triangleq [W_1, W_2, \ldots, W_N]$ is calculated by $W = P_0^{-1}H_0$.

Proof With the definition of $e_i(k) \triangleq z_i(k) - \hat{z}_i(k)$, the LE error can be calculated as:

$$e_i(k+1) = \mathcal{A}_i e_i(k) + \mathcal{M}_i \Delta a_i(k+1), \tag{5.12}$$

where $\mathcal{A}_i = \bar{A}_i - L_i\bar{C}_i\bar{A}_i$ and $\mathcal{M}_i = \bar{M}_i - L_i\bar{C}_i\bar{M}_i$. Then, considering the Lyapunov function $V_i(k) = e_i^T(k)P_ie_i(k)$, the deviation of $V_i(k)$ with respect to time k is

$$\begin{aligned} \Delta V_i(k) =& V_i(k+1) - V_i(k) \\ =& e_i^{\mathrm{T}}(k)(\mathcal{A}_i^{\mathrm{T}}P_i\mathcal{A}_i - P_i)e_i(k) + 2e_i^{\mathrm{T}}(k)\mathcal{A}_i^{\mathrm{T}}P_i\mathcal{M}_i\Delta a_i(k) \\ &+ \Delta a_i^{\mathrm{T}}(k)\mathcal{M}_i^{\mathrm{T}}P_i\mathcal{M}_i\Delta a_i(k). \end{aligned} \tag{5.13}$$

According to the definition of $\Delta a_i(k)$, it is obvious that $\Delta a_i(k)$ is energy-bounded. Hence, the classical H_∞ performance index [5] is introduced to maintain the satisfactory estimation performance of the intermediate estimator in case of attack variation.

$$J_i \triangleq \sum_{k=0}^{\infty}(e_i^{\mathrm{T}}(k)e_i(k) - \gamma_i^2\Delta a_i^{\mathrm{T}}(k)\Delta a_i(k)), \tag{5.14}$$

where γ_i is the H_∞ attack variation attenuation level bound. Under the zero-initial condition and $V_i(\infty) \geq 0$, it follows that

$$\begin{aligned} J_i \leq& \sum_{k=1}^{\infty}(e_i^{\mathrm{T}}(k)e_i(k) - \gamma_i^*\Delta a_i^{\mathrm{T}}(k)\Delta a_i(k)) + V_i(\infty) - V_i(0) \\ \leq& \sum_{k=1}^{\infty}(e_i^{\mathrm{T}}(k)e_i(k) - \gamma_i^*\Delta a_i^{\mathrm{T}}(k)\Delta a_i(k) + \Delta V_i(k)) \\ =& \eta_i^{\mathrm{T}}(k)\Xi_i\eta_i(k), \end{aligned} \tag{5.15}$$

where $\eta(k) = \left[e_i^{\mathrm{T}} \; \Delta a_i^{\mathrm{T}}(k)\right]^{\mathrm{T}}$, $\gamma_i^* = \gamma_i^2$, and

$$\Xi_i = \begin{bmatrix} \mathcal{A}_i^{\mathrm{T}}P_i\mathcal{A}_i - P_i + \mathbf{I} & \mathcal{A}_i^{\mathrm{T}}P_i\mathcal{M}_i \\ * & \mathcal{M}_i^{\mathrm{T}}P_i\mathcal{M}_i - \gamma_i^2\mathbf{I} \end{bmatrix}.$$

By letting $\Xi_i < 0$ and according to the Schur complement lemma [4], the inequality (5.15) is equivalent to (5.10). It is obvious that $J_i < 0$ holds for any $\eta_i(k) \neq 0$ and energy-bounded $\Delta a_i(k)$, which means that the local estimation error (5.12) is globally asymptotically stable.

Note that with redundant information from multiple sensors, estimation performance and reliability for system states and actuator attacks can be improved. Hence, to design the distributed secure fusion estimation algorithm, we have

$$\hat{\xi}(k) = \sum_{i=1}^{N} W_i \hat{\xi}_i(k), \tag{5.16}$$

and the fusion estimation error is calculated by:

$$e_0(k) \triangleq \xi(k) - \hat{\xi}(k) = \sum_{i=1}^{N} W_i e_{\xi i}(k), \tag{5.17}$$

where $e_{\xi i}(k) = \xi(k) - \hat{\xi}_i(k)$.

To determine a group of optimal weighting matrices W_i in (5.17), we derive the following fusion estimation error system according to (5.7), (5.9) and (5.17):

$$\begin{cases} e_F(k+1) = A_F e_F(k) + M_F e_a(k), \\ e_0(k) = W e_F(k), \end{cases} \tag{5.18}$$

where

$$\begin{cases} e_F(k) \triangleq \mathrm{col}\{e_{\xi 1}(k), \ldots, e_{\xi N}(k)\}, \\ e_a(k) \triangleq \mathrm{col}\{e_{a1}(k), \ldots, e_{aN}(k)\}, \\ e_{ai}(k) \triangleq \mathrm{col}\{e_{yi}(k), \Delta a_i(k+1)\}, \\ e_{yi}(k) \triangleq a_{yi}(k) - \hat{a}_{yi}(k). \end{cases}$$

Then, we can introduce the following Lyapunov function and class H_∞ performance index for (5.18):

$$\begin{cases} V(k) \triangleq e_0^{\mathrm{T}}(k) P_0 e_0(k), \\ J \triangleq \sum_{k=0}^{\infty} \left[e_0^{\mathrm{T}}(k) e_0(k) - \gamma^2 e_a^{\mathrm{T}}(k) e_a(k) \right]. \end{cases} \tag{5.19}$$

Due to the zero-initial condition and $V(\infty) \geq 0$, it follows that

$$
\begin{aligned}
J &\leq \sum_{k=1}^{\infty} \left[e_0^{\mathrm{T}}(k)e_0(k) - \gamma^2 e_a^{\mathrm{T}}(k)e_a(k)\right] + V(\infty) - V(0), \\
&\leq \sum_{k=1}^{\infty} \left[e_0^{\mathrm{T}}(k)e_0(k) - \gamma^2 e_a^{\mathrm{T}}(k)e_a(k) + \Delta V(k)\right] = \zeta^{\mathrm{T}}(k)\Xi\zeta(k),
\end{aligned}
$$

where $\Delta V(k) = V(k) - V(k-1)$, $\zeta(k) = \mathrm{col}\{e_F(k), e_a(k)\}$, and $\gamma^\star = \gamma^2$,

$$
\Xi = \begin{bmatrix} A_F^{\mathrm{T}} W^{\mathrm{T}} P_0 W A_F - P_0 + \mathbf{I} & A_F^{\mathrm{T}} W^{\mathrm{T}} P_0 W M_F \\ * & M_F \mathrm{T} W^{\mathrm{T}} P_0 W M_F - \gamma^\star \mathbf{I} \end{bmatrix}.
$$

According to the Schur complement lemma [4], $\Xi < 0$ is equivalent to the inequality in (5.11). Besides, it is obvious that $J < 0$ holds for any $\zeta(k) \neq 0$ and $e_a(k)$ when $\Xi < 0$, which means that the fusion estimation error system (5.18) is asymptotically stable. This completes the proof.

Based on Theorem 5.1, the algorithm of the proposed distributed secure fusion estimation for system (5.1) is summarized as follows:

Algorithm 5.1 Secure H_∞ Fusion Estimation

Input: $\mu, T, A, B, C_i, D_i \ (i = 1, 2, \cdots, N)$
Output: $\hat{\xi}(k), \hat{a}_{yi}(k)$
1: Calculate the local estimator gains L_i by (5.10);
2: Calculate the optimal weighting matrices W_i by (5.11);
3: Construct the local intermediate estimator by (5.9);
4: **for** $i = 1$ to N **do**
5: Calculate LE $\hat{z}_i(k)$ by (5.9);
6: Calculate SHFE $\hat{\xi}(k)$ by (5.16);
7: **end for**

Remark 5.2 It is important to highlight that the optimization problems delineated in (5.10) and (5.11) of Theorem 5.1 can be efficiently solved using the "*mincx*" function within MATLAB's LMI Toolbox [4]. A key advantage of this approach over online iterative optimization based estimation methods is that the computation of local estimator gains and optimal weighting matrices is conducted offline, which can enhance the scalability and ensures immediate responses, making it more practical for real-time applications.

5.4 Simulation

To validate the efficacy of our proposed secure fusion estimation algorithm, we utilize a practical scenario involving a maneuvering vehicle. This vehicle is closely monitored by a network of three sensors, as depicted in Fig. 5.1.

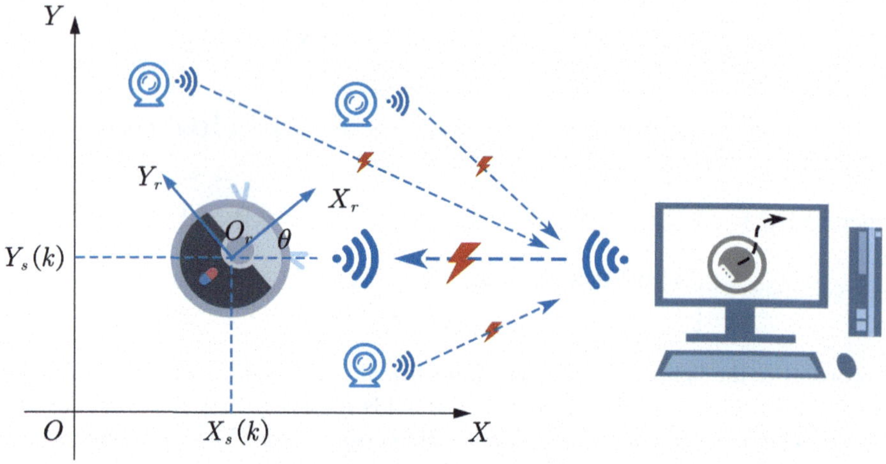

Fig. 5.1 The coordinate frame for the vehicle

The state vector is defined as $x(k) = \text{col}\{X_s(k), \dot{X}_s(k), Y_s(k), \dot{Y}_s(k)\}$, where $(X_s(k), Y_s(k))$ are the position coordinates along X- and Y-axes, respectively. Meanwhile, $(\dot{X}_s(k), \dot{Y}_s(k))$ are the corresponding velocities. Moreover, the system parameters are given as follows.

$$A = \begin{bmatrix} 1 & 0.5 & 0 & 0 \\ 0 & 1 & 0 & 0 \\ 0 & 0 & 1 & 0.5 \\ 0 & 0 & 0 & 1 \end{bmatrix}, B = \begin{bmatrix} 1 \\ 0 \\ 0 \\ 1 \end{bmatrix}, C_1 = \begin{bmatrix} 1 & 1 & 0 & 0 \\ 1 & 0 & 0 & 1 \\ 0 & 0 & 1 & 1 \end{bmatrix}, D_1 = \begin{bmatrix} 1 \\ 0 \\ 0 \end{bmatrix},$$

$$C_2 = \begin{bmatrix} 0 & 1 & 1 & 0 \\ 0 & 0 & 0 & 1 \\ 1 & 0 & 0 & 0 \end{bmatrix}, D_2 = \begin{bmatrix} 0 \\ 1 \\ 0 \end{bmatrix}, C_3 = \begin{bmatrix} 1 & 0 & 0 & 1 \\ 0 & 0 & 1 & 0 \\ 0 & 1 & 0 & 1 \end{bmatrix}, D_3 = \begin{bmatrix} 0 \\ 0 \\ 1 \end{bmatrix}.$$

Following Algorithm 5.1, we implement the secure H_∞ fusion estimation for system states and FDI attacks, using parameters $\mu = 0.7$ and $T = 50$. Figure 5.2a presents the trajectory of a maneuvering vehicle along the X- and Y-axes, illustrating the impact of FDI attacks on its movement compared with a non-attack scenario. This demonstrates the effectiveness of our proposed local intermediate estimator in accurately tracking the vehicle, even with corrupted actuator and sensor measurements. The enhanced trajectory estimation of our proposed SHFE algorithm is evident. Further, Fig. 5.2b showcases the actuator attack estimates from both local estimators and the FC, highlighting the satisfactory performance in FDI attack estimation. For sensor attacks, Figs. 5.3a–c display the local estimators' performance in estimating FDI attacks, under the condition that each sensor experiences distinct malicious attacks. Despite these challenges, the local intermediate estimators successfully reconstruct the FDI attacks on sensors. A comparative analysis, shown in Fig. 5.2d, uses mean

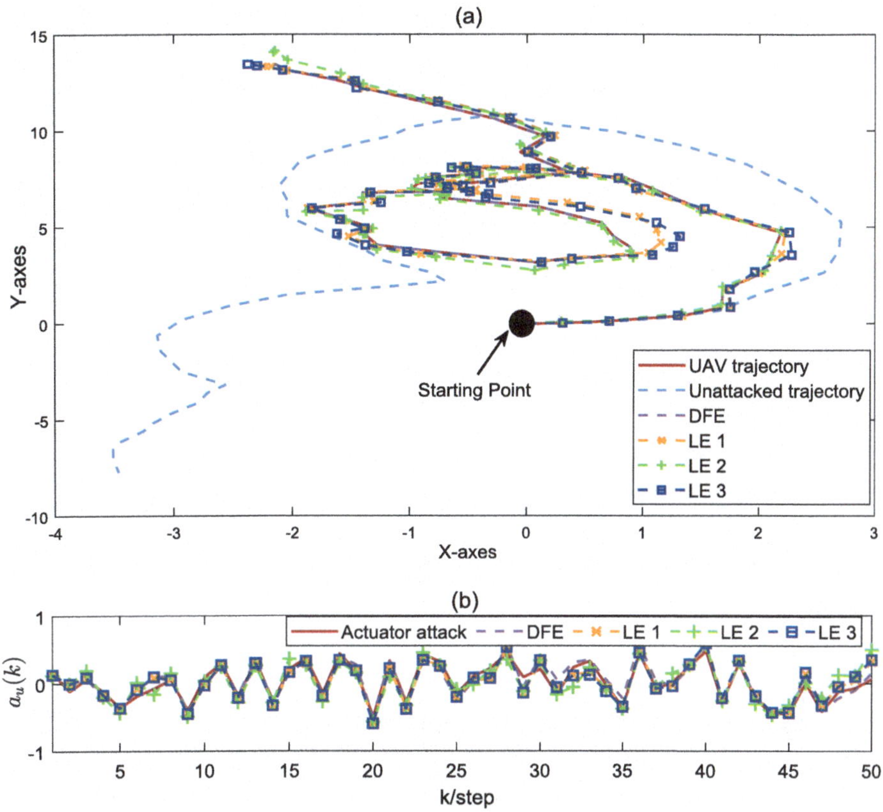

Fig. 5.2 **a** The vehicle's trajectory and its position estimation; **b** Actuator attack and its estimation

square errors (MSEs) derived from an averagely 100 Monte Carlo simulations to compare with the performance of LE and SHFE. The results indicate a significant improvement in LE performance when augmented by our proposed SHFE algorithm.

5.5 Conclusion

In this chapter, the secure H_∞ fusion estimation problem for linear discrete-time CPSs with FDI attacks was studied. By introducing an intermediate variable, the augmented system consists of system states, actuator attacks and local sensor attacks is modeled, and then local estimates of each sensor are generated by collecting its own measurements. By integrating the local estimates with our proposed optimal

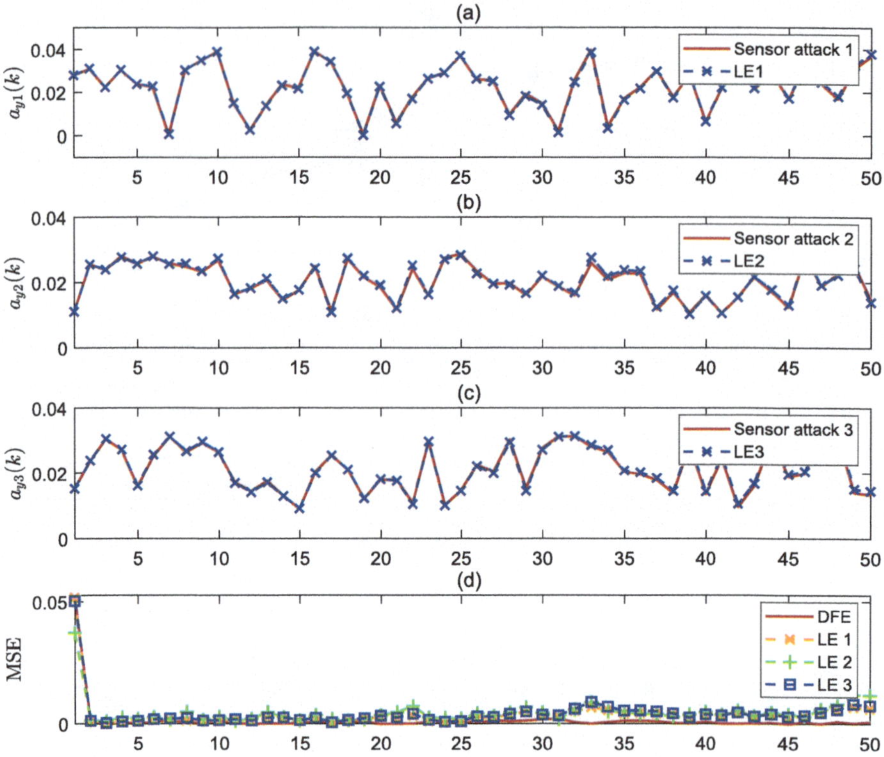

Fig. 5.3 **a–c** Senor attacks and attack estimates; **d** MSEs of the proposed SHFE and LEs

weighting fusion rules, the improved estimates of system states and actuator attacks can be obtained at the FC. Finally, the trajectory tracking simulation of a maneuvering vehicle is exploited to demonstrate the advantages and effectiveness of the proposed SHFE methods.

References

1. Lu AY, Yang GH (2018) Secure Luenberger-like observers for cyber-physical systems under sparse actuator and sensor attacks. Automatica 98:124–129
2. Zhou J, Chen B, Yu L (2020) Intermediate-variable-based estimation for FDI attacks in cyber-physical systems. IEEE Trans Circ Systems II: Express Briefs 67(11):2762–2766
3. Sun S, Lin H, Ma J, Li X (2017) Multi-sensor distributed fusion estimation with applications in networked systems: a review paper. Inf. Fusion 38:122–134
4. Boyd SP, Ghaoui LE, Feron E, Balakrishnan V (1994) Linear matrix inequalities in system and control theory. SIAM, Philadelphia, PA, USA
5. Liu Y, Arunkumar A, Sakthivel R et al (2020) Finite-time event-triggered non-fragile control and fault detection for switched networked systems with random packet losses. J Frankl Inst 357(16):11394–11420

Chapter 6
Distributed Fusion Estimation for Nonlinear CPSs with Attacked Control Signals

6.1 Introduction

In Chap. 5, intermediate-variable-based secure H_∞ fusion estimation method is designed against dual-channel FDI attacks. However, it only concerns the linear systems while the nonlinear systems are more flexible and popular in practical application. In this case, this chapter shall study the nonlinear secure fusion estimation methods for nonlinear CPSs. To capture the accurate dynamics with malicious information induced by FDI attacks, some works investigating the secure estimation for CPSs arose recently [1, 2]. For attack detection, traditional model-based methods always generate residuals to identify the occurrence of attacks. The performance indices can be divided into two categories, the robustness against disturbance [3, 4] and the sensitivity to attacks [5, 6], while attack estimation can also be taken into consideration. The false data injected at the actuator and sensors were estimated in [7] based on an intermediate-variable-based method, and [8] designed the distributed estimator for interconnected systems based on robust performance. In [9], an adaptive Kalman filter was developed for joint estimation of system parameter and state. Note that the methods above mostly focus on linear systems, while limited results are available for nonlinear systems, which involve more difficulties in analyzing the stability of designed secure estimation method. On the other hand, multi-sensor fusion estimation gained an increasing interest due to its high estimation accuracy and robustness [10]. It is noted that the performance of estimators should be ensured due to the resource limitation induced by limited energy and communication. Thus the distributed fusion structure, which is able to process information collaboratively and ensure the scalability of estimating methods, is a reliable approach to tackle redundant information from different sensors [11].

This chapter investigates the secure estimation problem for cyber-physical systems where the control input is falsified by FDI attack. First, novel optimization objectives are constructed and attack estimation is performed from real and expected measurement outputs, where the estimator gains are derived by solving quadratic problems recursively. Then, the exponential stability of the proposed estimator is

B. Chen et al., *Secure Fusion Estimation and Control for Cyber-Physical Systems Under Malicious Attacks*, https://doi.org/10.1007/978-981-97-9396-9_6

derived when the nonlinear observability rank condition is satisfied. Meanwhile, the distributed fusion criterion is designed and the weighting matrices are obtained to enhance estimation performance. Finally, an intelligent localization system monitored by multiple sensors is utilized to verify the effectiveness of the proposed method.

6.2 Problem Statement

Consider the nonlinear systems that are modeled by:

$$\begin{cases} x(t+1) = f(x(t)) + B(x(t))u(t) + E(t)w(t), \\ y_i(t) = h_i(x(t)) + E_i(t)v_i(t), i = 1, \cdots, L, \end{cases} \tag{6.1}$$

where $x(t) \in \Re^n$ and $y_i(t) \in \Re^{m_i}$ are the state and measurement of sensor i. $w(t) \in \Re^n$ and $v_i(t) \in \Re^{m_i}$ are bounded noises satisfying

$$\|w(t)\|_2 < \infty, \|v_i(t)\|_2 < \infty, \tag{6.2}$$

$E(t)$ and $E_i(t)$ are time-varying matrices with appropriate dimensions, and $E_i(t)$ is assumed to be full row rank. $f(x(t))$ and $h(x(t))$ are continuously differentiable nonlinear vector functions. When the control input $u(t) \in \Re^z$ is tampered with by FDI attack, the actual control input becomes:

$$\tilde{u}(t) = u(t) + a(t), \tag{6.3}$$

where $a(t)$ is FDI attack signal, and $\|a(t)\|_2 < \infty$ is also bounded.

Then, the local state estimator (LSE) $\hat{x}_i(t)$ is given as the following structure:

$$\begin{cases} \hat{x}_i^p(t) = f(\hat{x}_i(t-1)) + B_i(t-1)u(t-1), \\ r_i(t) = V_i(t)(y_i(t) - h_i(\hat{x}_i^p(t))), \\ \hat{x}_i^{p,u}(t) = \hat{x}_i^p(t) + B_i(t-1)r_i(t), \\ \hat{x}_i(t) = \hat{x}_i^{p,u}(t) + K_i(t)(y_i(t) - h_i(\hat{x}_i^{p,u}(t))), \end{cases} \tag{6.4}$$

where $B_i(t) = B_i(\hat{x}_i(t))$, $\hat{x}_i^p(t)$ is the one-step prediction, and $\hat{x}_i^{p,u}(t)$ represents the prediction with attack compensation. $r_i(t)$ is used to approach attack signal $a(t-1)$. $K_i(t)$ and $V_i(t)$ are the estimator gains to be designed.

With the designed LSEs $\hat{x}_i(t)$, the distributed fusion estimation (DFE) $\hat{x}_0(t)$ is given by:

$$\hat{x}_0(t) = \sum_{i=1}^{L} W_i(t)\hat{x}_i(t), \tag{6.5}$$

where $\sum_{i=1}^{L} W_i(t) = \mathbf{I}$.

Consequently, this chapter aims to achieve the following three targets:

1. Design $V_i(t)$ such that $r_i(t)$ can approach the attack signal $a(t)$.
2. Design $K_i(t)$ such that estimation error is exponentially bounded as $t \to \infty$.
3. Design optimal fusion weigting matrices $W_i(t)$, $(i = 1, 2, \cdots, L)$ such that the upper bound of fusion estimation error is minimum.

6.3 Main Result

6.3.1 Secure State Estimation

In this section, local estimator gain matrices $K_i(t)$, $V_i(t)$ and fusion weighting matrices $W_i(t)(i = 1, 2, \cdots, L)$ are designed. Before presenting the main result, define

$$\begin{cases} \tilde{x}_i(t) \triangleq x_i(t) - \hat{x}_i(t), \\ \tilde{r}_i(t) \triangleq r_i(t) - a(t-1), \\ A_i(t) \triangleq \frac{\partial f(x(t))}{\partial x(t)}\Big|_{x(t)=\hat{x}_i(t)}, \\ C_i(t) \triangleq \frac{\partial h_i(x(t))}{\partial x(t)}\Big|_{x(t)=\hat{x}_i^p(t)}, \\ \xi_i(t) \triangleq \text{col}\{a(t-1), \tilde{w}_i(t-1), \tilde{v}_i(t)\}, \\ \tilde{w}_i(t) \triangleq w(t) + O(\|\tilde{x}_i(t)\|^2), \\ \tilde{v}_i(t) \triangleq v_i(t) + O(\|x_i(t) - \hat{x}_i^p(t)\|^2), \end{cases} \tag{6.6}$$

where $\tilde{w}_i(t)$ and $\tilde{v}_i(t)$ are the combined effects of system disturbance and high order linearized error, respectively. Then, the following error dynamics are obtained

$$\begin{cases} \tilde{x}_i(t) \triangleq A_i^e(t)\tilde{x}_i(t-1) + B_i^e(t)\xi_i(t), \\ \tilde{r}_i(t) \triangleq C_i^e(t)\tilde{x}_i(t-1) + D_i^e(t)\xi_i(t), \end{cases} \tag{6.7}$$

where

$$\begin{cases} A_i^e(t) = G_i(t)F_i(t)A_i(t-1), \\ B_i^e(t) = \left[\mathcal{B}_{1_i}(t)\ \mathcal{B}_{2_i}(t)\ \mathcal{B}_{3_i}(t)\right], \\ C_i^e(t) = V_i(t)C_i(t)A_i(t-1), \\ D_i^e(t) = \left[\mathcal{D}_{1_i}(t)\ \mathcal{D}_{2_i}(t)\ \mathcal{D}_{3_i}(t)\right], \\ F_i(t) = \mathbf{I} - B_i(t-1)V_i(t)C_i(t), \\ G_i(t) = \mathbf{I} - K_i(t)C_i(t), \\ \mathcal{B}_{1_i}(t) = G_i(t)F_i(t)B_i(t-1), \\ \mathcal{B}_{2_i}(t) = G_i(t)F_i(t)E(t-1), \\ \mathcal{B}_{3_i}(t) = -(\mathbf{I} - K_i(t)C_i(t))B_i(t-1)V_i(t)E_i(t) - K_i(t)E_i(t), \\ \mathcal{D}_{1_i}(t) = V_i(t)C_i(t)B_i(t-1) - \mathbf{I}, \\ \mathcal{D}_{2_i}(t) = V_i(t)C_i(t)E(t-1), \mathcal{D}_{3_i}(t) = V_i(t)E_i(t). \end{cases} \tag{6.8}$$

To facilitate the analysis, the following matrices are further defined:

$$\begin{cases} A_{\dagger_i}(t) \triangleq A_i(t)A_i^{\mathrm{T}}(t), \\ B_{\dagger_i}(t) \triangleq B_i(t)B_i^{\mathrm{T}}(t), \\ E_{\dagger}(t) \triangleq E(t)E^{\mathrm{T}}(t), \\ E_{\dagger_i}(t) \triangleq E_i(t)E_i^{\mathrm{T}}(t), \\ \Lambda_{i,1}(t) \triangleq F_i(t)(A_{\dagger_i}(t-1) + B_{\dagger_i}(t-1) + E_{\dagger}(t-1))F_i^{\mathrm{T}}(t), \\ \Lambda_{i,2}(t) \triangleq B_i(t-1)V_i(t)E_{\dagger_i}(t)V_i^{\mathrm{T}}(t)B_i^{\mathrm{T}}(t-1), \\ A_f(t) \triangleq \mathrm{diag}\{A_1^e(t), \cdots, A_L^e(t)\}, \\ B_f(t) \triangleq \mathrm{diag}\{B_1^e(t), \cdots, B_L^e(t)\}, \\ \Omega(t) \triangleq [W_1(t), \cdots, W_{L-1}(t)], \\ \Psi_1 \triangleq \begin{bmatrix} \mathbf{I} & \mathbf{0} & \mathbf{0} & -\mathbf{I} \\ & & & \\ * & \ddots & \mathbf{0} & \vdots \\ * & * & \mathbf{I} & -\mathbf{I} \end{bmatrix} \in \mathfrak{R}^{(L-1)n\times Ln}, \\ \Psi_2 \triangleq [\mathbf{0} \cdots \mathbf{0}\ \mathbf{I}] \in \mathfrak{R}^{n\times Ln}, \\ \Lambda_f(t) \triangleq A_f(t)A_f^{\mathrm{T}}(t) + B_f(t)B_f^{\mathrm{T}}(t). \end{cases} \tag{6.9}$$

Theorem 6.1 Given positive scalars α and β, the local estimator gains $V_i(t)$ and $K_i(t)$ in (2.2) are given by

$$\begin{cases} V_i(t) = B_i^{\mathrm{T}}(t-1)C_i^{\mathrm{T}}(t)(\beta^2 E_{\dagger_i}(t) \\ \quad + \beta^2 C_i(t)(A_{\dagger_i}(t-1) + E_{\dagger}(t-1))C_i^{\mathrm{T}}(t) \\ \quad + C_i(t)B_{\dagger_i}(t-1)C_i^{\mathrm{T}}(t))^{-1}, \\ K_i(t) = (\Lambda_{i,1}(t) + \Lambda_{i,2}(t))C_i^{\mathrm{T}}(t)(C_i(t)(\Lambda_{i,1}(t) \\ \quad + \Lambda_{i,2}(t))C_i^{\mathrm{T}}(t) + \alpha^2 E_{\dagger_i}(t))^{-1}. \end{cases} \tag{6.10}$$

In this sense, the estimation performance of $\hat{x}_i(t)$ can be maintained in the presence of FDI attack $a(t)$. Moreover, the optimal fusion weighting matrices are given by

$$\begin{cases} \Omega(t) = -\Psi_2\Lambda_f^{\mathrm{T}}(t)\Psi_1^{\mathrm{T}}[\Psi_1\Lambda_f(t)\Psi_1^{\mathrm{T}}]^{-1}, \\ W_L(t) = \mathbf{I} - W_1(t) \cdots - W_{L-1}(t). \end{cases} \tag{6.11}$$

Proof First, we transform the structure of $\tilde{r}_i(t)$ by an introduced adjusting factor β, which leads to

$$\begin{aligned} \|\tilde{r}_i(t)\|_2 \le\ & \|[V_i(t)C_i(t)B_i(t-1) - \mathbf{I}]a(t-1)\|_2 \\ & + \|\beta V_i(t)C_i(t)A_i(t-1)\beta^{-1}\tilde{x}_i(t-1)\|_2 \\ & + \|\beta V_i(t)C_i(t)E(t-1)\beta^{-1}\tilde{w}(t-1)\|_2 \\ & + \|\beta V_i(t)E_i(t)\beta^{-1}\tilde{v}_i(t)\|_2. \end{aligned} \tag{6.12}$$

Note that

$$\begin{cases} \theta^{\mathrm{T}}\Gamma^{\mathrm{T}}\Gamma\theta \le \lambda_{\max}(\theta\theta^{\mathrm{T}})\mathrm{Tr}\{\Gamma^{\mathrm{T}}\Gamma\}, \\ \mathrm{Tr}\{\Delta\Gamma\} = \mathrm{Tr}\{\Gamma\Delta\}, \end{cases} \tag{6.13}$$

where θ is any vector and Δ and Γ are any matrices with appropriate dimensions. Then, it is derived that

$$\min Tr\{M_{\upsilon_i}(t)\} \tag{6.14}$$

can be regarded as the optimization problem for minimizing $\|\tilde{r}_i(t)\|_2$, where

$$\begin{aligned} M_{\upsilon_i}(t) &= \beta^2 V_i(t)C_i(t)A_i(t-1)(V_i(t)C_i(t)A_i(t-1))^{\mathrm{T}} \\ &+(V_i(t)C_i(t)B_i(t-1)-\mathbf{I})(V_i(t)C_i(t)B_i(t-1)-\mathbf{I})^{\mathrm{T}} \\ &+\beta^2 V_i(t)C_i(t)E(t-1)(V_i(t)C_i(t)E(t-1))^{\mathrm{T}} \\ &+\beta^2 V_i(t)E_i(t)(V_i(t)E_i(t))^{\mathrm{T}}. \end{aligned}$$

Reconstruct the optimization problem (6.14) as the following form

$$\min \theta^{\mathrm{T}} M_{\upsilon_i}(t)\theta. \tag{6.15}$$

By taking the partial derivative of the objective function in (6.15) with respect to $\theta^{\mathrm{T}} V_i(t)$ and setting it to zero, gain $V_i(t)$ can be obtained as (6.10). Moreover, by taking the second derivative, it can be obtain that

$$\frac{\partial^2 \theta^{\mathrm{T}} M_{\upsilon_i}(t)\theta}{\partial(\theta^{\mathrm{T}} V_i(t))^2} > 0. \tag{6.16}$$

This proves that $V_i(t)$ is the optimal solution to (6.15).

Subsequently, $K_i(t)$ can be also derived in a similar line with $V_i(t)$. But the difference lies in that an adjusting factor α is introduced and $K_i(t)E_i(t)\tilde{\upsilon}_i(t)$ in (6.7) is transformed into:

$$\alpha K_i(t)E_i(t)\alpha^{-1}\tilde{\upsilon}_i(t). \tag{6.17}$$

The detailed derivation is omitted for brevity.

On the other hand, the fusion estimation error is represented by following equation:

$$\tilde{x}_0(t) = W(t)A_f(t)\tilde{x}_f(t) + W(t)B_f(t)\xi_f(t), \tag{6.18}$$

where

$$\begin{cases} W(t) \triangleq [W_1(t), \cdots, W_L(t)], \\ \tilde{x}_f(t) \triangleq \mathrm{col}\{\tilde{x}_1(t), \cdots, \tilde{x}_L(t)\}, \\ \xi_f(t) \triangleq \mathrm{col}\{\xi_1(t), \cdots, \xi_L(t)\}, \end{cases} \tag{6.19}$$

and $A_f(t)$ and $B_f(t)$ are defined in (6.9). Similarly, the fusion weighting matrices $W_i(t)(i = 1, 2, \cdots, L)$ can be obtained by solving the following problem

$$\min\{\theta^{\mathrm{T}} M_f(t)\theta\}, \tag{6.20}$$

where $M_f(t) = W(t)[A_f(t)A_f^{\mathrm{T}}(t) + B_f(t)B_f^{\mathrm{T}}(t)]W^{\mathrm{T}}(t)$. Substituting $W(t) = \Omega(t)$ $\Psi_1 + \Psi_2$ into (6.20) and taking the partial derivative of the objective function in (6.20) with respect to $\theta^{\mathrm{T}}(t)\Omega(t)$, it yields that

$$\theta^{\mathrm{T}}\Omega(t)\Psi_1\Lambda_f(t)\Psi_1^{\mathrm{T}} + \theta^{\mathrm{T}}(t)\Psi_2\Lambda_f^{\mathrm{T}}(t)\Psi_1^{\mathrm{T}} = \mathbf{0}, \tag{6.21}$$

where Ψ_1 and Ψ_2 are defined in (6.9). If $\Lambda_f(t) \geq 0$, it is proved that the solution $\Omega(t)$ given in (6.11) is the optimal fusion weighting matrix. This completes the proof.

Remark 6.1 The principle to derive $V_i(t)$ is eliminating the impact of attack signal on state estimation, rather than minimizing the attack estimation error $\tilde{r}_i(t)$. Except the solution in (6.10), $V_i(t)$ can be obtained by minimizing $\| V_i(t)C_i(t)B_i(t-1) - \mathbf{I}\|_2$, and it can be derived that

$$V_i(t) = \begin{cases} \Im^{\mathrm{T}}(t)(\Im(t)\Im^{\mathrm{T}}(t))^{-1}, & \Im(t) \text{ is full row rank,} \\ (\Im^{\mathrm{T}}(t)\Im(t))^{-1}\Im^{\mathrm{T}}(t), & \Im(t) \text{ is full column rank,} \end{cases}$$

where $\Im(t) = C_i(t)B_i(t-1)$. Notice that the rank of $\Im(t)$ may be undetermined due to the time-varying matrices. Hence the above solution is insoluble under some cases. To avoid such a limitation, a scalar $0 < \beta \leq 1$ is introduced. Decreasing β appropriately, it can not only obtain a more reliable estimation of $a(t)$, but also avoid the deterioration in extreme circumstances, such as $\|C_i(t)\| \to 0$.

Remark 6.2 If the attack signal $a(t)$ has a slow change rate, the estimation performance of $r_i(t)$ can be improved by utilizing recursive least square (RLS) algorithm. The recursion for the new estimator $\hat{r}_i(t)$ is defined as follows. Given forgetting factor λ and initial matrix $\mathcal{S}_i(0)$, we have

$$r_i^p(t) = \mathcal{C}_i(t+1)\hat{r}_i(t), \tag{6.22}$$
$$\hat{r}_i(t+1) = \hat{r}_i(t) + \mathcal{S}_i(t)\mathcal{C}_i(t)\mathcal{P}_i(t)(r(t+1) - r_i^p(t)), \tag{6.23}$$

where $r_i^p(t)$ is one-step prediction, $\hat{r}_i(0) = r_i(0)$ and

$$\begin{cases} \mathcal{C}_i(t) = V_i(t)C_i(t)B_i(t-1), \\ \mathcal{P}_i(t) = (\lambda\mathbf{I} + \mathcal{C}_i(t)\mathcal{S}_i(t)\mathcal{C}_i^{\mathrm{T}}(t))^{-1}, \\ \mathcal{S}_i(t+1) = \frac{1}{\lambda}\mathcal{S}_i(t) - \frac{1}{\lambda}\mathcal{S}_i(t)\mathcal{C}_i^{\mathrm{T}}(t)\mathcal{P}_i(t)\mathcal{C}_i(t)\mathcal{S}_i(t). \end{cases}$$

6.3.2 Stability Analysis

This section derives the estimator gain for attack-free case and provides the boundedness analysis of estimation error in a comprehensible way. The nonlinear system under the attack-free case is modeled as follows:

$$\begin{cases} x(t+1) = f(x(t)) + E(t)w(t), \\ y(t) = h(x(t)) + E_1(t)v(t), \end{cases} \tag{6.24}$$

where the system parameters are consistent with (6.1). Then, the exponential boundedness analysis of estimation error $\tilde{x}_i(t)$ in the presence of FDI attack is obtained in a similar line. In the meantime, the role that α plays is revealed. Before deriving the main result, the following assumptions and lemma are introduced.

Assumption 6.1 The system state $x(t)$ and the measurement function $h(x(t))$ are bounded at each time instant.

Assumption 6.2 Suppose that the Jacobian matrices of nonlinear functions $f(x(t))$ and $h(x(t))$ are $A(t)$ and $C(t)$, respectively. Then, there exist positive real numbers $\bar{a}, \underline{a}, \bar{c}, \underline{c}, \bar{e}, \underline{e} > 0$, such that $\underline{a} \le \|A(t)\| \le \bar{a}$, $\underline{c} \le \|C(t)\| \le \bar{c}$, and $\underline{e} \le \|E_1(t)\| \le \bar{e}$ are valid at each time instant.

Lemma 6.1 [12]: Assume there is a stochastic process $\mathcal{V}(\tilde{x}(t))$ as well as real numbers $\underline{v}, \bar{v}, \mu > 0$ and $0 < \kappa \le 1$ such that

$$\underline{v}\,\|\tilde{x}(t)\|^2 \le \mathcal{V}(\tilde{x}(t)) \le \bar{v}\,\|\tilde{x}(t)\|^2$$

and

$$\mathbb{E}\{\mathcal{V}(\tilde{x}(t+1))|\mathcal{V}(\tilde{x}(t))\} - \mathcal{V}(\tilde{x}(t)) \le \mu - \kappa\mathcal{V}(\tilde{x}(t))$$

are valid for every solution of $\tilde{x}(t)$. Then, $\tilde{x}(t)$ is exponentially bounded in mean square, i.e.,

$$\mathbb{E}\{\|\tilde{x}(t)\|^2\} \le \frac{\bar{v}}{\underline{v}}\mathbb{E}\{\|\tilde{x}(0)\|^2\}(1-\kappa)^t + \frac{\mu}{\underline{v}}\sum_{i=1}^{t-1}(1-\kappa)^i.$$

Theorem 6.2 For nonlinear system (6.24), the estimator $\hat{x}(t)$ is given as the following structure

$$\hat{x}(t) = f(\hat{x}(t-1)) + K(t)(y(t) - h(f(\hat{x}(t-1)))), \tag{6.25}$$

where the estimator gain $K(t)$ is give by

$$K(t) = \Phi(t)C^{\mathrm{T}}(t)(C(t)\Phi(t)C^{\mathrm{T}}(t) + \alpha^2\mathcal{E}(t))^{-1} \tag{6.26}$$

and α is a given scalar, $\Phi(t) = A(t-1)A^{\mathrm{T}}(t-1) + E(t-1)E^{\mathrm{T}}(t-1)$ and $\mathcal{E}(t) = E_1(t)E_1^{\mathrm{T}}(t)$. Under Assumptions 6.1 and 6.2, when $C(t)$ is full column rank and initial error $\tilde{x}(0) \triangleq x(0) - \hat{x}(0)$ is bounded, estimation error $\tilde{x}(t) \triangleq x(t) - \hat{x}(t)$ is exponentially bounded if α satisfies

$$\begin{cases} 0 < \alpha, & \delta \leq 0, \\ 0 < \alpha < (1-\kappa)^{\frac{1}{4}} \underline{c}^{\frac{3}{2}} \underline{a} (\bar{a}\bar{c}\delta)^{-\frac{1}{2}}, & \delta > 0, \end{cases} \tag{6.27}$$

where $\delta = \bar{e}^2 - \sqrt{1-\kappa}\underline{e}^2\underline{c}(\bar{a}\bar{c})^{-1}$ and $0 < \kappa \leq 1$.

Proof Based on Eqs. (6.24) and (6.25), the estimation error $\tilde{x}(t)$ is given by

$$\begin{aligned} \tilde{x}(t) = {} & G(t)A(t-1)\tilde{x}(t-1) \\ & + G(t)E(t-1)\tilde{w}(t-1) - \alpha K(t)E_1(t)\alpha^{-1}\tilde{v}(t), \end{aligned} \tag{6.28}$$

where $G(t) = \mathbf{I} - K(t)C(t)$. Under Assumption 6.1, the innovation error $y(t) - h(f(\hat{x}(t-1)))$ is bounded. Then it can be concluded that $\tilde{x}(t)$ is bounded at each time instant. Meanwhile, disturbances $\tilde{w}(t)$ and $\tilde{v}(t)$ in (6.28) combined with additional linearization error are bounded. In this case, the estimator gain $K(t)$ can be obtained in a similar line as Theorem 6.1.

Subsequently, introduce a stochastic process

$$\mathcal{V}(t) = \tilde{x}^{\mathrm{T}}(t)\wp(t)\tilde{x}(t), \tag{6.29}$$

where $\wp(t) = C^{\mathrm{T}}(t)C(t)$. Notice that $\wp(t) > 0$ if $C(t)$ is of full column rank and there exist real numbers $\underline{v}$ and $\bar{v}$ that satisfy $\underline{v}\mathbf{I} < \wp(t) < \bar{v}\mathbf{I}$. To analyze the exponential boundedness of estimation error $\tilde{x}(t)$ by Lemma 6.1, by pre-multiplying $G(t)$ by $C(t)$, it is obtained that $C(t)G(t) = \aleph(t)C(t)$, where

$$\aleph(t) = \alpha^2\mathcal{E}(t)(C(t)\Phi(t)C^{\mathrm{T}}(t) + \alpha^2\mathcal{E}(t))^{-1} \tag{6.30}$$

and $\|\aleph(t)\|$ can approach 0 while α approaches 0. It follows that there exist scalars $\alpha > 0$ and $0 < \kappa \leq 1$ such that

$$\begin{aligned} A^{\mathrm{T}}(t)C^{\mathrm{T}}(t+1)\aleph^{\mathrm{T}}(t+1)\aleph(t+1)C(t+1)A(t) \\ - C^{\mathrm{T}}(t)C(t) \leq -\kappa C^{\mathrm{T}}(t)C(t). \end{aligned} \tag{6.31}$$

Obviously, if (6.31) is valid, it can be obtained that

$$\begin{aligned} & \mathcal{V}(t+1) - \mathcal{V}(t) \leq -\kappa\mathcal{V}(t) \\ & + \xi^{\mathrm{T}}(t+1)G_{\xi}^{\mathrm{T}}(t+1)\wp(t+1)G_{\xi}(t+1)\xi(t+1) \\ & + 2\xi^{\mathrm{T}}(t+1)G_{\xi}^{\mathrm{T}}(t+1)\wp(t+1)G(t+1)A(t)\tilde{x}(t), \end{aligned} \tag{6.32}$$

where $G_{\xi}(t) = [G(t)E(t-1), K(t)E_1(t)]$ and $\xi(t) = \text{col}\{\tilde{w}(t-1), \tilde{v}(t)\}$. Then, $\tilde{x}(t)$ is exponentially bounded in mean square if initial estimation error $\tilde{x}(0)$ is bounded.

Moreover, the range of α can be described through $\aleph(t)$ in light of (6.31) and Assumption 6.2. On the one hand, inequality (6.31) is strictly valid if

$$\aleph^{\mathrm{T}}(t+1)\aleph(t+1) \leq (1-\kappa)\frac{\underline{c}^2}{\bar{a}^2\bar{c}^2}\mathbf{I}. \tag{6.33}$$

On the other hand, according to (6.30), the upper bound of $\aleph^{\mathrm{T}}(t)\aleph(t)$ is constructed as follows:

$$\aleph^{\mathrm{T}}(t)\aleph(t) \leq \alpha^4\bar{e}^4(\alpha^2\underline{e}^2 + \underline{c}^2\underline{a}^2)^{-2}\mathbf{I}. \tag{6.34}$$

Combine (6.34) and (6.33), then it is derived that (6.31) can be ensured if the following inequality holds.

$$\alpha^2\bar{e}^2(\alpha^2\underline{e}^2 + \underline{c}^2\underline{a}^2)^{-1} \leq \sqrt{(1-\kappa)}\frac{\underline{c}}{\bar{a}\bar{c}}. \tag{6.35}$$

As a result, the range of α in (6.27) is obtained. It should be pointed out that the range of α is sufficient for the boundedness of $\tilde{x}(t)$.

Remark 6.3 Theorem 6.2 is under the premise that $\wp(t) > 0$, which is limited because $C(t)$ may not have full column rank. To generalize the results to a more general situation, based on Eqs. 4.1 to 4.6 from [13], a observability rank condition is further investigated, which is

$$\underline{v}\mathbf{I} \leq \sum_{j=t-M}^{t} \Phi^{\mathrm{T}}(j,t)C^{\mathrm{T}}(t)C(t)\Phi(j,t) \leq \bar{v}\mathbf{I}, \tag{6.36}$$

where $\Phi(j,t) = A(t-1)\cdots A(j)$ and $\Phi(t,t) = \mathbf{I}$. It turns out that if the system satisfies nonlinear observability rank condition, there exist a stochastic process and a scalar α for estimator gain $K(t)$ that match Lemma 6.1, and an appropriate α is closely associated with the exponential boundedness of $\tilde{x}(t)$. Furthermore, the convergence rate of estimation error $\tilde{x}(t)$ can be adjusted through α. The smaller α is, the greater κ can be and the faster $\tilde{x}(t)$ converges.

Remark 6.4 The upper bound of α can be derived by the procedure above. However, in this chapter, we prefer to determine α by simulation result in advance, which can find a suitable value of α and greatly reduce the amount of online computation.

Remark 6.5 Given estimator (6.4) for nonlinear system (6.1) in the presence of FDI attack, if nonlinear observability rank condition is satisfied, there exists an estimator gain $K_i(t)$ with a positive scalar α ensuring the exponential boundedness of estimation error $\tilde{x}_i(t)$.

Proof According to Remark 6.3, introduce the following stochastic process

$$\mathcal{V}_i(t) = \tilde{x}_i(t)\wp_i(t)\tilde{x}_i^{\mathrm{T}}(t), \tag{6.37}$$

where $\wp_i(t) = \sum_{j=t-M}^{t} \Phi_i^{\mathrm{T}}(t)(j,t)C_i^{\mathrm{T}}(t)C_i(t)\Phi_i(j,t) > 0$. Since the form of $K_i(t)$ and $\tilde{x}_i(t)$ are similar to (6.26) and (6.28), respectively, it is able to derive that the range of α lies in $0 < \alpha < \bar{\alpha}$, and the upper bound $\bar{\alpha}$ can be obtained in a similar line with Theorem 6.2. The detailed proof is omitted here.

6.4 Simulation

Consider an intelligent localization system for the target unicycle-like mobile robot with two driving wheels, in which the common axis passes through the center of the robot. Denote $S_x(t)$ and $S_y(t)$ as the coordinates of the midpoint between the two driving wheels, and $\theta(t)$ as the angle between the main axis of the robot and the x-direction. The control inputs $u_p(t)$ and $u_r(t)$ respectively control translation speed and rotational speed. Given sampling period T_0, the system is described by the following discretized model [14, 15]:

$$\begin{cases} S_x(t+1) = S_x(t) + T_0 u_p(t)\cos\theta(t), \\ S_y(t+1) = S_y(t) + T_0 u_p(t)\sin\theta(t), \\ \theta(t+1) = \theta(t) + T_0 u_r(t). \end{cases} \tag{6.38}$$

Define state $x(t) \triangleq \mathrm{col}\{S_x(t), S_y(t), \theta(t)\}$ and attack signal $a(t) \triangleq \mathrm{col}\{a_p(t), a_r(t)\}$. Then, the robot system (6.38) under the bounded process noise $w(t)$ and the FDI attack $a(t)$ injected in the actuator can be modeled as

$$x(t+1) = f(x(t)) + B(x(t))a(t) + E(t)w(t), \tag{6.39}$$

where $E(t) = 0.3\mathbf{I}$ and

$$\begin{cases} f(x_i(t)) = \begin{bmatrix} S_x(t) + T_0 u_p(t)\cos\theta(t) \\ S_y(t) + T_0 u_p(t)\sin\theta(t) \\ \theta(t) + T_0 u_r(t) \end{bmatrix}, \\ B(x(t)) = \begin{bmatrix} T_0\cos\theta(t) & \mathbf{0} \\ T_0\sin\theta(t) & \mathbf{0} \\ \mathbf{0} & T_0 \end{bmatrix}. \end{cases} \tag{6.40}$$

Suppose that there are six sensors located at $(S_{x_j}, S_{y_j})(j = 1, \cdots, 6)$ to measure the distance $d_j(t)$ and the azimuth $\varphi_j(t)$

$$\begin{cases} d_j(t) = \sqrt{(S_{x_j} - S_x(t))^2 + (S_{y_j} - S_y(t))^2}, \\ \varphi_j(t) = \theta(t) - \arctan\left\{\frac{S_{y_j} - S_y(t)}{S_{x_j} - S_x(t)}\right\}. \end{cases} \tag{6.41}$$

In the simulation, the sensors are divided into three groups for distributed fusion. Then, the measurement model under the bounded noise $v_i(t)$ is expressed as

$$y_i(t) = h_i(x(t)) + E_i(t)v_i(t), i = 1, 2, 3, \tag{6.42}$$

where $E_1(t) = E_2(t) = E_3(t) = \mathbf{I}$ and

$$h_i(x(t)) = \text{col}\{d_{i_1}(t), \varphi_{i_1}(t), d_{i_2}(t), \varphi_{i_2}(t)\}. \tag{6.43}$$

In this case, the linearized jacobian matrices $A_i(t)$ and $C_i(t)$ of $f(x(t))$ and $h_i(x(t))$ are given by

$$\begin{cases} A_i(t) = \begin{bmatrix} 1 & 0 & -T_0 u_p(t)\sin\theta(t) \\ 0 & 1 & T_0 u_p(t)\cos\theta(t) \\ 0 & 0 & 1 \end{bmatrix}_{x(t)=\hat{x}_i(t)}, \\ C_i(t) = \text{col}\{C_i^j(t), C_i^{j+1}(t)\}, \; j = 2i - 1, \end{cases} \tag{6.44}$$

where $\hat{x}_i(t) \triangleq \text{col}\{S_{\hat{x}_i}(t), S_{\hat{y}_i}(t), \hat{\theta}(t)\}$ is the state estimation, $\hat{x}_i^p(t)$ is the corresponding one-step prediction, and

$$\begin{cases} C_i^j(t) = \begin{bmatrix} \frac{-S_{\tilde{x}_j}(t)}{\sqrt{S_{xy}(t)}} & \frac{-S_{\tilde{y}_j}(t)}{\sqrt{S_{xy}(t)}} & 0 \\ \frac{-S_{\tilde{y}_j}(t)}{S_{xy}(t)} & \frac{S_{\tilde{y}_j}(t)}{S_{xy}(t)} & 1 \end{bmatrix}_{x(t)=\hat{x}_i^p(t)}, \\ S_{\tilde{x}_j}(t) = S_{x_j} - S_{\hat{x}_j}(t), \\ S_{\tilde{y}_j}(t) = S_{y_j} - S_{\hat{y}_j}(t), \\ S_{xy}(t) = S_{\tilde{x}_j}^2(t) + S_{\tilde{y}_j}^2(t). \end{cases} \tag{6.45}$$

In this example, sampling period is set as $T_0 = 1s$, and the control inputs are chosen as $u_p(t) = 0.3$ and $u_r(t) = 0.15$. The initial value of state $\hat{x}_i(0)$ is set as the actual state $x(0) = \text{col}\{0.1, 0.2, 1\}$. Due to the particularity of the function $\arctan(\cdot)$, the positions of six sensors divided into three groups are set as $\{(-20, -20), (20, -20)\}$, $\{(-15, 15), (15, 15)\}$, $\{(-20, 20), (20, 20)\}$ and the continuity of $\arctan(\cdot)$ can be guaranteed. The robot moves within a certain range and will not exceed 90° of the sensor's viewing angle starting from x-direction or y-direction for sure. Moreover, the bounded noises $w(t)$ and $v_i(t)$ are given as

$$\begin{cases} w(t) = \text{col}\{0.9\rho_{x_1}(t) - 0.5, 0.7\rho_{x_2}(t) - 0.4, \\ \qquad 0.8\rho_{x_3}(t) - 0.5\}, \\ v_i(t) = \text{col}\{0.12\rho_{y_i^1}(t) - 0.07, 0.1\rho_{y_i^2}(t) - 0.05, \\ \qquad 0.11\rho_{y_i^3}(t) - 0.06, 0.09\rho_{y_i^4}(t) - 0.06\}, \end{cases} \tag{6.46}$$

where $\rho_{x_i}(t) \in [0, 1]$ and $\rho_{y_i^j}(t) \in [0, 1]$ $(i = 1, 2, 3, j = 1, 2, 3, 4)$ are random variables and are generated by "rand" function in MATLAB. For further analysis, define $\rho(t) \in [0, 1]$ similarly with $\rho_{x_i}(t)$ and the following two types of attack signals $a(t)$ will be considered:

1. **Type I**:

$$\begin{cases} a_p(t) = \begin{cases} 0.8\cos(t/15) & 50 < t < 200, \\ & \text{else}, \end{cases} \\ (t) = 0. \end{cases}$$

2. **Type II**:

$$\begin{cases} a_p(t) = \begin{cases} 0.5(\rho(t) - 0.5) & 50 \le t < 150, \\ & \text{else}, \end{cases} \\ (t) = \begin{cases} 0.3(\rho(t) - 0.5) & 50 \le t < 150, \\ & \text{else}. \end{cases} \end{cases}$$

The scalars in Theorem 6.1 are set as $\beta = 0.1$ and $\alpha = 1$. First, under Type I of FDI attacks, the robot's trajectories are shown in Fig. 6.1, and it is intuitive that the red line obtained based on Theorem 6.1 is better than the blue line obtained based on adaptive extended Kalman filtering (AEKF) from [9], where covariances are unknown and inaccurate covariances are used. Meanwhile, the attack signal estimation is compared

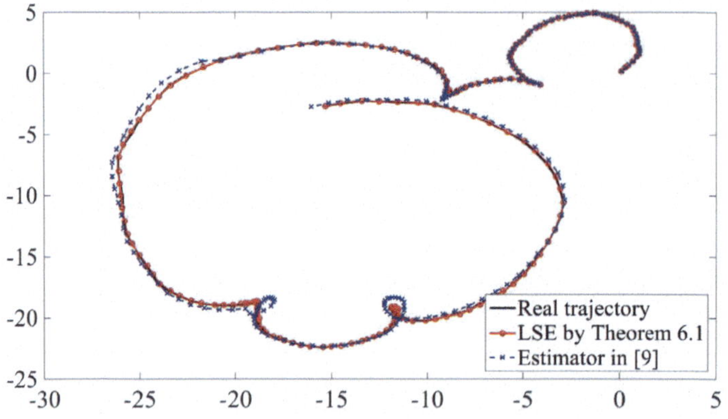

Fig. 6.1 The real trajectory of the robot under FDI attack of Type I

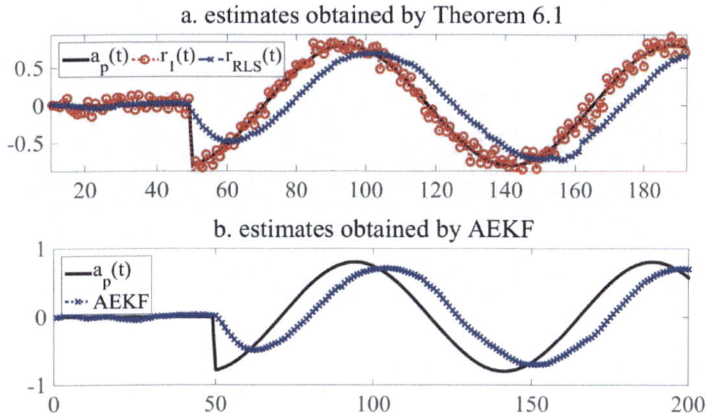

Fig. 6.2 Attack estimation under different methods

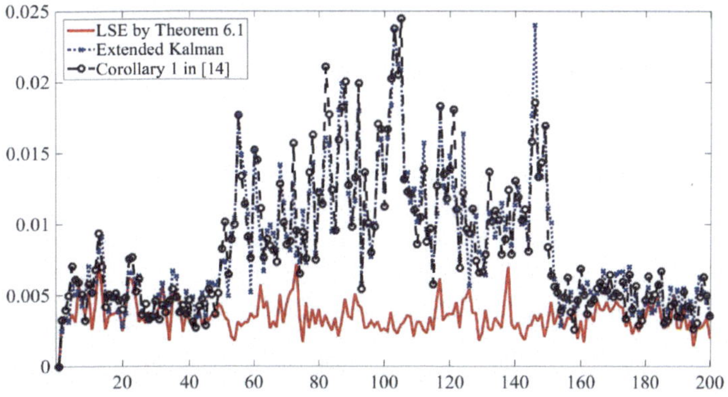

Fig. 6.3 The comparison of estimation performance under different nonlinear filtering methods

in Fig. 6.2. In Fig. 6.2a, the red line is obtained by $r_i(t)$ which is generated by the estimator (6.4), and the blue line is the improved estimator proposed in Remark 6.2. It can been found that, after implementing RLS, the obtained curve is lagging but more smooth, and it shows a similar property with the curve in Fig. 6.2b obtained by AEKF.

Besides, under Type II of FDI attacks, the proposed estimator is compared with other nonlinear methods in Fig. 6.3. By carrying out Monte Carlo simulations with 50 runs in average, the result shows that the method proposed in this chapter is better than EKF and the filter obtained by solving linear matrix inequality recursively, whether there is an attack or not [14]. Finally, effectiveness of the fusion approach is illustrated by Fig. 6.4, and the result shows that the performance of DFE has better robustness than LSEs.

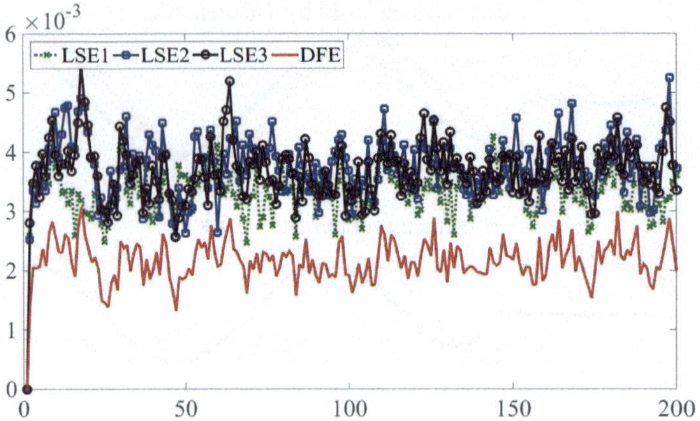

Fig. 6.4 Comparison of the DFE and LSEs

6.5 Conclusion

The distributed fusion estimation problem where FDI attack occurs at the actuator was addressed in this chapter. Through first-order Taylor series expansion, the problem was handled in terms of linear time-varying sense. Firstly, a secure estimator was designed with control input compensation, and the estimator gains were derived by solving quadratic problem recursively. Secondly, performance of local estimators can be enhanced in terms of the fusion criteria. Notice that analytical solutions for the parameters of desired estimators were derived for the purpose of online operation. Finally, an illustrative example was employed to demonstrate the effectiveness of the proposed nonlinear estimator.

References

1. Ding D, Han QL, Ge X, Wang J (2021) Secure state estimation and control of cyber-physical systems: a survey. IEEE Trans Syst Man, Cybernet Syst 51(1):176–190
2. An L, Yang GH (2019) Distributed secure state estimation for cyber-physical systems under sensor attacks. Automatica 107:526–538
3. Li Y, Zhong M (2011) Fault detection filter design for linear discrete time-varying systems with multiplicative noise. J Syst Eng Electron 22(6):982–990
4. Zhong M, Zhou D, Ding SX (2010) On designing $\mathcal{H}_\infty$ fault detection filter for linear discrete time-varying systems. IEEE Trans Automat Cont 55(7):1689–1695
5. Zhong M, Ding SX, Ding EL (2010) Optimal fault detection for linear discrete time-varying systems. Automatica 46(8):1395–1400
6. Zhong M, Zhang L, Ding SX, Zhou D (2017) A probabilistic approach to robust fault detection for a class of nonlinear systems. IEEE Trans Ind Electron 64(5):3930–3939
7. Zhou J, Chen B, Yu L (2020) Intermediate-variable-based estimation for FDI attacks in cyber-physical systems. IEEE Trans Circ Syst. II, Express Briefs, 67(11):2762–2766

8. Zhang K, Jiang B, Shi P (2019) Distributed fault estimation observer design with adjustable parameters for a class of nonlinear interconnected systems. IEEE Trans Cybernet 49(12):4219–4228
9. Zhang Q (2018) Adaptive Kalman filter for actuator fault diagnosis. Automatica 93:333–342
10. Li XR, Zhu Y, Wang J, Han C (2003) Optimal linear estimation fusion. I. Unified fusion rules. IEEE Trans Inf Theo 49(9):2192–2208
11. Chen B, Hu G, Ho DWC, Yu L (2019) A new approach to linear/nonlinear distributed fusion estimation problem. IEEE Trans Autom Cont 64(3):1301–1308
12. Reif K, Gunther S, Yaz E, Unbehauen R (1999) Stochastic stability of the discrete-time extended Kalman filter. IEEE Trans Autom Cont 44(4):714–728
13. Song Y, Grizzle JW (1992) The extended Kalman filter as a local asymptotic observer for nonlinear discrete-time systems. 1992 American Control Conference
14. Chen B, Hu G (2018) Nonlinear state estimation under bounded noises. Automatica 98:159–168
15. Jetto L, Longhi S, Venturini G (1999) Development and experimental validation of an adaptive extended Kalman filter for the localization of mobile robots. IEEE Trans Rob Automat 15(2):219–229

Chapter 7
Networked Nonlinear Fusion Estimation Under DoS Attacks

7.1 Introduction

In Chaps. 2 and 3, efficient secure fusion estimation methods based on data privacy protection strategies were proposed from the perspective of proactive defense. Chapter 4 proposed a fusion estimation method to improve the speed of attack detection and warning. From the perspective of passive defense, Chaps. 5 and 6 developed several secure fusion estimation methods to defend against FDI attacks. In this chapter, a bounded recursive optimization approach is adopted to defend against the unknown DoS attacks. Since the DoS attack is the easiest way to access the attack space and block the communication channels [1], some researchers are motivated to contrive assorted networked fusion estimation methods for cyber security. Generally, most security fusion estimation methods are based on Kalman filter [2–5]. For instance, a dimensionality reduction approach was developed to defend against DoS attacks and bandwidth constraints in [3] based on Kalman fusion estimation. In [4], the DoS attacks were modeled as the case of measurement packet dropout, and a bank of Kalman filters were proposed to detect and isolate the possible attacks. Moreover, based on the idea of data block approach, multi-rate Kalman filters were designed in [5] at each state updating instant.

Though various secure fusion estimation methods have been developed to counter DoS attacks, they heavily rely on the quantity and distribution of the attack signals. However, the statistical characteristics of attack signals and system noises are too difficult to be accurately obtained, even not obeying any random distributions. Meanwhile, most cyber-physical systems are nonlinear in the actual environment, but only a few works have focused on distributed nonlinear fusion estimation under DoS attacks. In this case, the secure fusion estimation for the nonlinear networked multi-sensor system under DoS attacks still faces some challenges: (1) How to design the nonlinear estimator to estimate system state while maintaining satisfactory estimation performance under DoS attack? (2) How to design a fusion algorithm without the knowledge of the noise statistical characteristics and attack signal distributions?

B. Chen et al., *Secure Fusion Estimation and Control for Cyber-Physical Systems Under Malicious Attacks*,https://doi.org/10.1007/978-981-97-9396-9_7

Motivated by analysis above, a nonlinear fusion estimation method against DoS attacks will be proposed in this chapter, and a target tracking system is also given to verify the effectiveness of the proposed method.

7.2 Problem Statement

Consider a discrete physical process monitored by L sensors (see Fig. 7.1), and the nonlinear state-space equation can be modeled by

$$\mathrm{x}(t+1) = f(\mathrm{x(t)}) + B(t)w(t), \tag{7.1}$$

where $\mathrm{x(t)} \in \Re^n$ is the system state, $f(\mathrm{x(t)}) \in \Re^{n\times 1}$ is a continuously differentiable nonlinear function, $B(t)$ is a known time-varying matrix with appropriate dimension, and $w(t) \in \Re^n$ is the process noise.

It can be seen from Fig. 7.1 that L sensors are employed for monitoring the system state (7.1), then each sensor measurement equation is given by

$$z_i(t) = g_i(\mathrm{x(t)}) + D_i(t)v_i(t), i \in \{1, 2, \cdots, L\}, \tag{7.2}$$

where $g_i(\mathrm{x(t)}) \in \Re^{m_i\times 1}$ is a continuously differentiable nonlinear function, $D_i(t)$ is a known time-varying matrix with appropriate dimension, and $v_i(t) \in \Re^{m_i}$ is the measurement noise. Particularly, $w(t)$ and $v_i(t)$ are both bounded noises and satisfy

$$w^T(t)\,w(t) \le \delta_w,\ {v_i}^T(t)\,v_i(t) \le \delta_{v_i}, \tag{7.3}$$

where δ_w and δ_{v_i} are unknown.

As shown in Fig. 7.1, when the measurement $z_i(t)$ is sent to the corresponding local estimator over communication networks, an adversary can launch DoS attacks to block the communication channels. In this case, the local estimators may not receive the measurement information $z_i(t)$ at time t. To describe this phenomenon clearly, a binary variable $\gamma_i(t)$ is introduced as follows:

$$\gamma_i(t) = \begin{cases} 0 & \text{if a DoS attack is launched,} \\ 1 & \text{otherwise.} \end{cases}$$

Notice that the successful attacks by adversary will cause a reduction in the amount of measurements received by the estimator, and then degrade the estimation accuracy of the estimator. To ensure that the designed estimator still maintains satisfactory estimation performance, a measurement compensation strategy is presented to compensate the attacked measurements. Specifically, a compensation measurement model can be given by

$$y_i(t) = \gamma_i(t)z_i(t) + (1 - \gamma_i(t))y_i(t-1), i = 1, 2, ..., L, \tag{7.4}$$

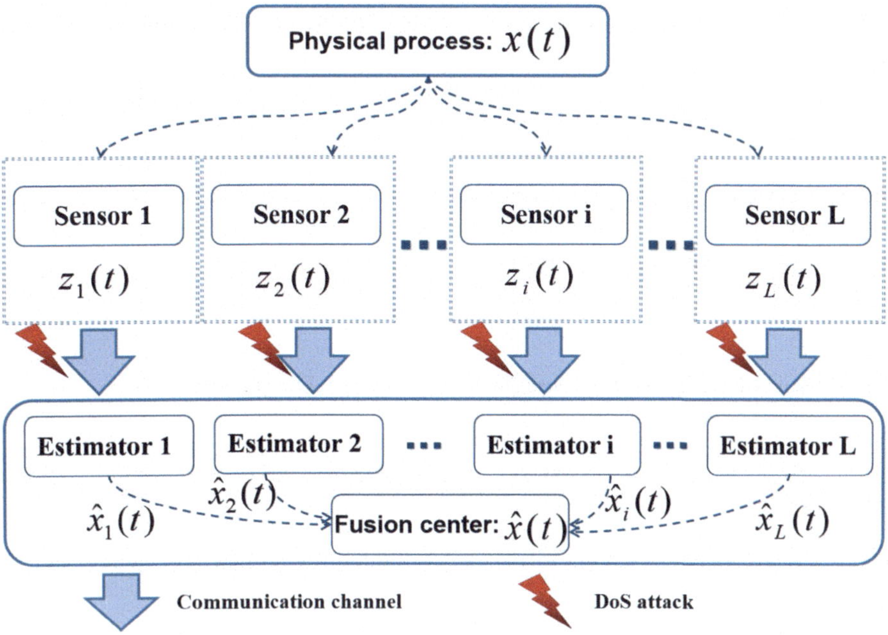

Fig. 7.1 The distributed secure fusion estimation structure for a class of nonlinear networked multi-sensor systems under DoS attacks

where $y_i(t)$ is the received measurement by the local estimator. Then, the measurement information $\{y_i(1), y_i(2), \ldots, y_i(t)\}$ will be used to design local nonlinear estimator.

Firstly, define $X_i(t) = \left[\mathrm{x}^\top(t), y_i^\top(t-1)\right]^\top$, $W_i(t) = \left[w^\top(t), v_i^\top(t)\right]^\top$, and combine (7.1)–(7.4), then an augmented system is constructed as:

$$X_i(t+1) = F_i(X_i(t)) + B_i(t)W_i(t), \tag{7.5}$$

$$y_i(t) = G_i(X_i(t)) + \gamma_i(t)D_i(t)v_i(t), i = 1, 2, \ldots, L, \tag{7.6}$$

where

$$F_i(X_i(t)) = \begin{bmatrix} f(\mathrm{x(t)}) \\ \gamma_i(t)g_i(\mathrm{x(t)}) \end{bmatrix} + \begin{bmatrix} \mathbf{0} & \mathbf{0} \\ \mathbf{0} & (1-\gamma_i(t))\mathbf{I} \end{bmatrix} X_i(t), \tag{7.7}$$

$$G_i(X_i(t)) = \gamma_i(t)g_i(\mathrm{x(t)}) + \left[\mathbf{0}\ (1-\gamma_i(t))\mathbf{I}\right] X_i(t), \tag{7.8}$$

$$B_i(t) = \begin{bmatrix} B(t) & \mathbf{0} \\ \mathbf{0} & \gamma_i(t)D_i(t) \end{bmatrix}. \tag{7.9}$$

Consider the augmented system (7.5), (7.6), the local state estimator (LSE) is designed by

$$\begin{cases} \hat{X}_i^{\mathrm{p}}(t) = F_i(\hat{X}_i(t-1)), \\ \hat{X}_i(t) = \hat{X}_i^{\mathrm{p}}(t) + \mathrm{K}_i(t)[y_i(t) - G_i(\hat{X}_i^{\mathrm{p}}(t))]. \end{cases} \tag{7.10}$$

where $\hat{X}_i(t)$ and $\hat{X}_i^{\mathrm{p}}(t)$ are the local estimate and one-step prediction of the state $X_i(t)$, respectively, and $\mathrm{K}_i(t)$ is the estimator gain to be determined.

Next, let $\hat{x}_i(t)$ denote the local estimate of $\mathrm{x}(t)$, then one has

$$\hat{\mathrm{x}}_i(t) = [\mathbf{I} \;\; \mathbf{0}]\hat{X}_i(t). \tag{7.11}$$

On the basis of $\hat{x}_i(t)$, the distributed fusion estimator (DFE) is given by

$$\hat{\mathrm{x}}(\mathrm{t}) = \sum_{i=1}^{L} \Omega_i(t)\hat{\mathrm{x}}_i(t), \tag{7.12}$$

where $\Omega_1(t)(i = 1, \ldots, L)$ are distributed weighted fusion matrices, which satisfy $\sum_{i=1}^{L} \Omega_i(t) = \mathbf{I}$.

Consequently, the main aims to be addressed in this chapter are to design nonlinear estimator gain $\mathrm{K}_i(t)$ in (7.10) and weighted fusion matrix $\Omega_i(t)$ in (7.12) such that the upper bound of square error of designed estimators is minimal at each time.

Remark 7.1: It can be concluded from the compensation model (7.4) and (7.5)–(7.10) that the raw measurement $z_i(t)$ can be received by the local estimator when $\gamma_i(t) = 1$, which means that the attack is not successful on the i-th channel at time t. If the adversary successively launches a DoS attack (i.e., $\gamma_i(t) = 0$), the local estimator may not receive a new measurement at that time, and then the latest measurement $y_i(t-1)$ will be used to compensate the destroyed measurement. In this case, the system (7.1)–(7.4) can be described in a compact form (7.5), (7.6) including the information of DoS attacks by augmentation.

7.3 Main Result

Local estimator gain $\mathrm{K}_i(t)$ and weighted fusion matrix $\Omega_i(t)$ will be determined by constructing two different convex optimization problems in this section. Define

$$
\begin{cases}
\mathrm{K}_{\mathrm{C}_i}(t) \triangleq \mathbf{I} - \mathrm{K}_i(t) C_{G_i}(t), \\
\mathrm{K}_{\mathrm{B}_i}(t) \triangleq \mathrm{K}_{\mathrm{C}_i}(t) B_i(t-1), \\
\mathrm{K}_{\mathrm{D}_i}(t) \triangleq -\gamma_i(t) \mathrm{K}_i(t) D_i(t), \\
\mathrm{B}_{\mathrm{D}_i}(t) \triangleq [\, \mathrm{K}_{\mathrm{B}_i}(t) \;\; \mathrm{K}_{\mathrm{D}_i}(t) \,], \\
\mathrm{A}_{\mathrm{F}}(t) \triangleq \mathrm{diag}\left\{ \mathrm{K}_{\mathrm{C}_1}(t) A_{F_1}(t-1), ..., \mathrm{K}_{\mathrm{C}_L}(t) A_{F_L}(t-1) \right\}, \\
\mathrm{B}_{\mathrm{F}}(t) \triangleq \mathrm{col}\left\{ \mathrm{B}_{\mathrm{D}_1}(t), ..., \mathrm{B}_{\mathrm{D}_L}(t) \right\}, \\
\Omega(t) \triangleq \left[\Omega_1(t) \; \mathbf{0} \; \Omega_2(t) \; \mathbf{0} \; ... \; \Omega_L(t) \; \mathbf{0} \right].
\end{cases}
\tag{7.13}
$$

Theorem 7.1 *For the nonlinear systems* (7.1), (7.2), *the local estimator gain* $K_i(t)$ *in (7.10) can be obtained by solving the following convex optimization problem:*

$$
\begin{aligned}
&\min_{\Pi_i(t)>0, \Upsilon_i(t)>0, M_i(t)>0, \eta_i(t)>0, \Psi_i(t), K_i(t)} \quad Tr\{\Upsilon_i(t)\} + Tr\{M_i(t)\} \\
&\text{s.t.} \begin{cases}
\begin{bmatrix}
-\mathbf{I} & K_{C_i}(t) A_{F_i}(t-1) & K_{B_i}(t) & K_{D_i}(t) \\
* & -\Pi_i(t) & \mathbf{0} & \mathbf{0} \\
* & * & -\Upsilon_i(t) & -\Psi_i(t) \\
* & * & * & -\Phi_i(t)
\end{bmatrix} < 0, \\
\Pi_i(t) - \eta_i(t)\mathbf{I} < 0, \\
\eta_i(t) \le \zeta_i,
\end{cases}
\end{aligned}
\tag{7.14}
$$

where $\zeta_i > 0$ *is a known scalar. Then, the distributed weighted fusion matrices* Ω_i $(i = 1, \ldots, L)$ *can be determined by solving the following convex optimization problem:*

$$
\begin{aligned}
&\min_{\Pi(t)>0, M(t)>0, \Psi(t), \Omega(t)} \quad Tr\{\Pi(t)\} + Tr\{M(t)\} \\
&\text{s.t.} \begin{bmatrix}
-\mathbf{I} & \Omega(t)\mathrm{A}_{\mathrm{F}}(t) & \Omega(t)\mathrm{B}_{\mathrm{F}}(t) \\
* & -\Pi(t) & -\Psi(t) \\
* & * & -\Phi(t)
\end{bmatrix} < 0.
\end{aligned}
\tag{7.15}
$$

Proof By defining prediction error $\tilde{X}_i^{\mathrm{p}}(t) = X_i(t) - \hat{X}_i^{\mathrm{p}}(t)$, then from (7.5) and (7.10) it yields that

$$
\tilde{X}_i^{\mathrm{p}}(t) \triangleq F_i(X(t-1)) - F_i(\hat{X}_i(t-1)) + B_i(t-1) W_i(t-1). \tag{7.16}
$$

Then, define local estimation error $\tilde{X}_i(t) \triangleq X_i(t) - \hat{X}_i(t)$, from (7.5) and (7.10), one has

$$
\tilde{X}_i(t) = \tilde{X}_i^{\mathrm{p}}(t) - \mathrm{K}_i(t)[\mathrm{G}_i(X_i(t)) - \mathrm{G}_i(\hat{X}_i^{\mathrm{p}}(t)) + \gamma_i(t) D_i(t) v_i(t)]. \tag{7.17}
$$

Utilize the first-order Taylor series expansion to linearize the nonlinear functions $\mathrm{F}_i(X_i(t-1))$ and $G_i(X_i(t))$ in (7.16) and (7.17), respectively:

$$\begin{cases} \mathrm{F}_i(X_i(t-1)) = \mathrm{F}_i(\hat{X}_i(t-1)) + A_{F_i}(t-1)\tilde{X}_i(t-1) + \Delta_{\mathrm{F}_i}(\tilde{X}_i^2(t-1)), \\ \mathrm{G}_i(X_i(t)) = \mathrm{G}_i(\hat{X}_i^{\mathrm{p}}(t)) + C_{\mathrm{G}_i}(t)\tilde{X}_i^{\mathrm{p}}(t) + \Delta_{\mathrm{G}_i}([\tilde{X}_i^{\mathrm{p}}(t)]^2), \end{cases} \tag{7.18}$$

where

$$\begin{cases} A_{F_i}(t-1) = \left.\frac{\partial \mathrm{F}_i(X_i(t-1))}{\partial X_i(t-1)}\right|_{X_i(t-1)=\hat{X}_i(t-1)}, \\ C_{G_i}(t) = \left.\frac{\partial \mathrm{G}_i(X_i(t))}{\partial X_i(t)}\right|_{X_i(t)=\hat{X}_i^{\mathrm{p}}(t)}, \end{cases} \tag{7.19}$$

and $\Delta_{\mathrm{F}_i}(\tilde{X}_i^2(t-1))$ and $\Delta_{\mathrm{G}_i}([\tilde{X}_i^{\mathrm{p}}(t)]^2)$ are the high-order terms of the Taylor expansion. Then, the nonlinear estimation error (7.17) can be rewritten as

$$\begin{aligned} \tilde{X}_i(t) = {} & \mathrm{K}_{\mathrm{C}_i}(t)A_{F_i}(t-1)\tilde{X}_i(t-1) \\ & + \mathrm{K}_{\mathrm{C}_i}(t)[B_i(t-1)W_i(t-1) + \Delta_{\mathrm{F}_i}(\tilde{X}_i^2(t-1))] \\ & - \mathrm{K}_i(t)[\gamma_i(t)D_i(t)v_i(t) + \Delta_{\mathrm{G}_i}([\tilde{X}_i^{\mathrm{p}}(t)]^2)]. \end{aligned} \tag{7.20}$$

Inspired by the idea of bounded recursive optimization in [7], the linearization errors $\Delta_{\mathrm{F}_i}(\tilde{X}_i^2(t-1))$ and $\Delta_{\mathrm{G}_i}([\tilde{X}_i^{\mathrm{p}}(t)]^2)$ can be regarded as bounded uncertainties. Meanwhile, it can be seen from (7.3) that $W_i(t-1)$ and $v_i(t)$ are unknown bounded noises. Thus (7.20) is written as

$$\begin{aligned} \tilde{X}_i(t) = {} & \mathrm{K}_{\mathrm{C}_i}(t)A_{F_i}(t-1)\tilde{X}_i(t-1) + \mathrm{K}_{\mathrm{C}_i}(t)B_i(t-1)\tilde{W}_i(t-1) \\ & - \gamma_i(t)\mathrm{K}_i(t)D_i(t)\tilde{v}_i(t), \end{aligned} \tag{7.21}$$

where $\tilde{W}_i(t-1)$ and $\tilde{v}_i(t)$ are unknown disturbances to model the influence factors caused by unknown noises and linearization errors.

To construct an upper bound of the square error of the LSE (7.10), the following performance index [7] is introduced:

$$\begin{aligned} \mathcal{J}_i(t) \triangleq {} & \tilde{X}_i^{\top}(t)\tilde{X}_i(t) - \tilde{X}_i^{\top}(t-1)\Pi_i(t)\tilde{X}_i(t-1) - \tilde{W}_i^{\top}(t-1)\Upsilon_i(t)\tilde{W}_i(t-1) \\ & - \tilde{v}_i^{\top}(t)\Phi_i(t)\tilde{v}_i(t) - \tilde{W}_i^{\top}(t-1)\Psi_i(t)\tilde{v}_i(t), \end{aligned} \tag{7.22}$$

where $\Pi_i(t) > 0$, $\Upsilon_i(t) > 0$, $\Phi_i(t) > 0$ and $\Psi_i(t)$ are introduced matrices. Then, substituting (7.21) into (7.22) yields

$$\mathcal{J}_i(t) \triangleq \begin{bmatrix} \tilde{X}_i(t-1) \\ \tilde{W}_i(t-1) \\ \tilde{v}_i(t) \end{bmatrix}^{\top} \underbrace{\begin{bmatrix} \Delta_i^1(t) & \Delta_i^2(t) & \Delta_i^3(t) \\ * & \Delta_i^4(t) & \Delta_i^5(t) \\ * & * & \Delta_i^6(t) \end{bmatrix}}_{\Delta_i(t)} \begin{bmatrix} \tilde{X}_i(t-1) \\ \tilde{W}_i(t-1) \\ \tilde{v}_i(t) \end{bmatrix}, \tag{7.23}$$

where

$$
\begin{cases}
\Delta_i^1(t) \triangleq A_{F_i}^{\top}(t-1)\mathrm{K}_{\mathrm{C}_i}^{\top}(t)\mathrm{K}_{\mathrm{C}_i}(t)A_{F_i}(t-1) - \Pi_i(t),\\
\Delta_i^2(t) \triangleq A_{F_i}^{\top}(t-1)\mathrm{K}_{\mathrm{C}_i}^{\top}(t)\mathrm{K}_{\mathrm{C}_i}(t)B_i(t-1),\\
\Delta_i^3(t) \triangleq A_{F_i}^{\top}(t-1)\mathrm{K}_{\mathrm{C}_i}^{\top}(t)\gamma_i(t)\mathrm{K}_i(t)D_i(t),\\
\Delta_i^4(t) \triangleq B_i^{\top}(t-1)\mathrm{K}_{\mathrm{C}_i}^{\top}(t)\mathrm{K}_{\mathrm{C}_i}(t)B_i(t-1) - \Upsilon_i(t),\\
\Delta_i^5(t) \triangleq B_i^{\top}(t-1)\mathrm{K}_{\mathrm{C}_i}^{\top}(t)\gamma_i(t)\mathrm{K}_i(t)D_i(t) - \Psi_i(t),\\
\Delta_i^6(t) \triangleq \gamma_i(t)D_i^{\top}(t)\mathrm{K}_i^{\top}(t)\gamma_i(t)\mathrm{K}_i(t)D_i(t) - \Phi_i(t).
\end{cases}
\tag{7.24}
$$

It can be deduced from (7.22) that $\mathcal{J}_i(t) < 0$ guarantees that $\tilde{X}_i^{\top}(t)\tilde{X}_i(t)$ has an upper bound, that is $\Delta_i(t) < 0$. Then, by using Schur complement lemma, $\Delta_i(t) < 0$ holds is equivalent to that the first inequality in (7.14) holds. Moreover, the second and third inequalities in (7.14) can be deduced by a similar derivation of Theorem 1 in [7].

From (7.22), the following inequality also holds:

$$
\begin{aligned}
\tilde{X}_i^{\top}(t)\tilde{X}_i(t) &< \eta_i(t)\tilde{X}_i^{\top}(t-1)\tilde{X}_i(t-1)\\
&\quad + \lambda_{\max}\left(\begin{bmatrix}\tilde{W}_i(t-1)\\ \tilde{v}_i(t)\end{bmatrix}\begin{bmatrix}\tilde{W}_i(t-1)\\ \tilde{v}_i(t)\end{bmatrix}^{\top}\right) Tr\left\{\begin{bmatrix}\Upsilon_i(t) & \Psi_i(t)\\ * & \Phi_i(t)\end{bmatrix}\right\},
\end{aligned}
\tag{7.25}
$$

where $\eta_i(t)$ has been defined in (7.14). Therefore, "$\min(Tr\{\Upsilon_i(t)\} + Tr\{\Phi_i(t)\})$" can be the optimization problem to minimize the upper bound of $\tilde{X}_i^{\top}(t)\tilde{X}_i(t)$, which can determine the LSE gain $\mathrm{K}_i(t)$ in (7.10).

Subsequently, the weighted fusion matrices in (7.15) can be determined. Define $e_i(t) = [\mathbf{I}\ \ \mathbf{0}]\tilde{X}_i(t)$ and $e(t) = \mathrm{x}(t) - \hat{\mathrm{x}}(t)$, then one has

$$
e(t) = \sum_{i=1}^{L}\left[\,\Omega_i(t)\ \mathbf{0}\,\right]\tilde{X}_i(t). \tag{7.26}
$$

It follows from (7.12) and (7.21) that

$$
\begin{cases}
e_F(t) = \mathrm{A_F}(t)\,e_F(t-1) + \mathrm{B_F}(t)\,\xi(t-1),\\
e(t) = \Omega(t)\,e_F(t),
\end{cases}
\tag{7.27}
$$

where $e_F(t) \triangleq \mathrm{col}\{e_1(t), \ldots, e_L(t)\}, \xi(t) \triangleq \mathrm{col}\{w(t-1), \mathbf{0}, v_1(t), \ldots, v_L(t)\}$.

Similarly, the square error of the fusion estimation $\hat{\mathrm{x}}(t)$ can be constructed by

$$
e^{\top}(t)\,e(t) < \begin{bmatrix}e_F(t-1)\\ \xi(t-1)\end{bmatrix}^{\top}\begin{bmatrix}\Pi(t) & \Psi(t)\\ * & \Phi(t)\end{bmatrix}\begin{bmatrix}e_F(t-1)\\ \xi(t-1)\end{bmatrix}, \tag{7.28}
$$

where $\Pi(t) > 0$, $\Phi(t) > 0$ and $\Psi(t)$ are introduced unknown matrices. Then, "$\min(Tr\{\Pi(t)\} + Tr\{\Phi(t)\})$" can be chosen as the optimization objective to establish a convex optimal problem for calculating the weighted matrices such that the upper bound of $e^{\top}(t)\,e(t)$ is minimal. The detailed derivation is omitted here.

Based on Theorem 7.1, the calculation process of fusion estimation $\hat{\mathrm{x}}(t)$ in (7.12) can be summarized as the following algorithm.

Algorithm 7.1 Distributed Secure Fusion Estimation under DoS Attacks

1: Initialization states $\mathrm{x}(0)$ and $\hat{\mathrm{x}}(t)$;
2: Calculate the matrices $A_{J_i}(t-1)$ and $C_{J_i}(t)$ in (7.19);
3: Solve the convex optimization problem (7.14) to determine local estimator gain $\mathrm{K}_i(t)$;
4: Calculate local estimation $\hat{\mathrm{x}}_i(t)$ by (7.10) and (7.11);
5: Solve the convex optimization problem (7.15) to determine weighting fusion matrix $\Omega_i(t)$;
6: Calculate the fusion estimation $\hat{\mathrm{x}}(t)$ by (7.12);
7: Return to Step 2 and implement Steps 2–6 for calculating $\hat{\mathrm{x}}(t+1)$.

Remark 7.2: Notice that Theorem 7.1 contains two different convex optimization problems with linear matrix inequality constraints, and then the local estimator gain $\mathrm{K}_i(t)$ and distributed weighting fusion matrix $\Omega_i(t)$ can be obtained by the function "mincx" in Matlab LMI Toolbox. Moreover, Algorithm 7.1 is implemented by the recursive form, and thus only the initial value $\hat{\mathrm{x}}_i(0)$ needs to be known in advance for calculating fusion estimation $\hat{\mathrm{x}}(t)$.

7.4 Simulation

Consider an intelligent vehicle tracking system in plane coordinates. Let $\mathrm{x}(t) \triangleq \mathrm{col}\{s_x(t), s_y(t), \theta(t)\}$ denote the vehicle's motion state, where $s_x(t)$ and $s_y(t)$ are the positions of the vehicle along X and Y axes, respectively, and $\theta(t)$ is the heading angle. When the vehicle moves along a circular trajectory, the motion model of the vehicle can be given by [8]

$$\begin{cases} s_x(t+1) = s_x(t) + \frac{\hat{u}_p}{\hat{u}_r}\cos\left(\theta(t) + T_0\hat{u}_r/2\right), \\ s_y(t+1) = s_y(t) + \frac{\hat{u}_p}{\hat{u}_r}\sin\left(\theta(t) + T_0\hat{u}_r/2\right), \\ \theta(t+1) = \theta(t) + T_0\hat{u}_r, \\ \hat{u}_p = u_p + w_p(t), \\ \hat{u}_r = u_r + w_r(t), \end{cases} \tag{7.29}$$

where u_p and u_r are translational velocity and rotational velocity control commands, respectively. In fact, these commands inevitable are subject to some external disturbances, i.e., $w_p(t)$ and $w_r(t)$ are unknown noises. Then, the state space equation (7.29) can be written as

$$\mathrm{x}(t+1) = f(\mathrm{x}(t)) + Bw(t), \tag{7.30}$$

where

$$\begin{cases} f(\mathrm{x}(t)) = \begin{bmatrix} s_x(t) + \frac{u_p}{u_r}\cos(\theta(t) + T_0 u_r/2) \\ s_y(t) + \frac{u_p}{u_r}\sin(\theta(t) + T_0 u_r/2) \\ \theta(t) + T_0 u_r \end{bmatrix}, \\ B \triangleq \mathrm{diag}\{1, 1, T_0\}, \\ w(t) = \mathrm{col}\left\{-\frac{u_p}{u_r}\cos(\theta(t) + T_0 u_r/2) + \frac{\hat{u}_p}{\hat{u}_r}\cos\left(\theta(t) + T_0\hat{u}_r/2\right), \right. \\ \quad -\frac{u_p}{u_r}\sin(\theta(t) + T_0 u_r/2) + \frac{\hat{u}_p}{\hat{u}_r}\sin\left(\theta(t) + T_0\hat{u}_r/2\right), \\ \quad \left. T_0 w_r(t)\right\}. \end{cases} \tag{7.31}$$

Moreover, four sensors are arranged in $(s_{x_{11}}, s_{y_{11}}) = (5, 10)$, $(s_{x_{12}}, s_{y_{12}}) = (5, 5)$, $(s_{x_{21}}, s_{x_{21}}) = (10, 5)$ and $(s_{x_{22}}, s_{x_{22}}) = (10, 10)$ to observe the vehicle's information. In this example, the distance and the angle can be measured from the sensors, then the measurement equation can be modeled by

$$\begin{cases} d_{ij}(t) = \sqrt{\left(s_{x_{ij}} - s_x(t)\right)^2 + \left(s_{y_{ij}} - s_y(t)\right)^2}, \\ \varphi_{ij}(t) = \theta(t) - \arctan\left(\frac{s_{y_{ij}} - s_y(t)}{s_{x_{ij}} - s_x(t)}\right), \end{cases} \tag{7.32}$$

where $(s_{x_{ij}}, s_{y_{ij}})(i = 1, 2, j = 1, 2)$ are the sensors' coordinates, and $(s_x(t), s_y(t))$ is the vehicle's coordinate in X-Y plane.

Therefore, the measurement equation can be described by

$$y_i(t) = \mathrm{g}_i(\mathrm{x}(t)) + D_i v_i(t), \tag{7.33}$$

where $\mathrm{g}_i(\mathrm{x}(t)) = \mathrm{col}\{d_{i1}(t), \varphi_{i1}(t), d_{i2}(t), \varphi_{i2}(t)\}$, and $v_i(t)(i = 1, 2)$ are the measurement noises.

Then, by using first-order Taylor series expansions to linearize the nonlinear functions $f(\mathrm{x}(t))$ and $\mathrm{g}_i(\mathrm{x}(t))$, the linearized matrices $A_{f_i}(t)$ and $C_{g_i}(t)(i = 1, 2)$ are calculated by

$$\begin{cases} A_{f_i}(t) = \begin{bmatrix} 1 & 0 & -\frac{u_p}{u_r}\sin(\theta(t) + T_0 u_r/2) \\ 0 & 1 & \frac{u_p}{u_r}\cos(\theta(t) + T_0 u_r/2) \\ 0 & 0 & 1 \end{bmatrix}_{\mathrm{x}(t)=\mathrm{x}^*}, \\ C_{g_i}(t) = \begin{bmatrix} \frac{-\tilde{s}_{x_{i1}}(t)}{\sqrt{\tilde{s}^2_{x_{i1}}(t)+\tilde{s}^2_{y_{i1}}(t)}} & \frac{-\tilde{s}_{y_{i1}}(t)}{\sqrt{\tilde{s}^2_{x_{i1}}(t)+\tilde{s}^2_{y_{i1}}(t)}} & 0 \\ \frac{-\tilde{s}_{y_{i1}}(t)}{\tilde{s}^2_{x_{i1}}(t)+\tilde{s}^2_{y_{i1}}(t)} & \frac{\tilde{s}_{x_{i1}}(t)}{\tilde{s}^2_{x_{i1}}(t)+\tilde{s}^2_{y_{i1}}(t)} & 1 \\ \frac{-\tilde{s}_{x_{i2}}(t)}{\sqrt{\tilde{s}^2_{x_{i2}}(t)+\tilde{s}^2_{y_{i2}}(t)}} & \frac{-\tilde{s}_{y_{i2}}(t)}{\sqrt{\tilde{s}^2_{x_{i2}}(t)+\tilde{s}^2_{y_{i2}}(t)}} & 0 \\ \frac{-\tilde{s}_{y_{i2}}(t)}{\tilde{s}^2_{x_{i2}}(t)+\tilde{s}^2_{y_{i2}}(t)} & \frac{\tilde{s}_{x_{i2}}(t)}{\tilde{s}^2_{x_{i2}}(t)+\tilde{s}^2_{y_{i2}}(t)} & 1 \end{bmatrix}_{\mathrm{x}(t)=\mathrm{x}^*}, \end{cases} \tag{7.34}$$

where $\tilde{s}_{x_{ij}}(t) \triangleq s_{x_{ij}}(t) - s_{x(t)}$ and $\tilde{s}_{y_{ij}}(t) \triangleq s_{y_{ij}}(t) - s_{y(t)}$. Since the communication channels may be subject to DoS attacks, the augmentation system can be moleded by (7.5)-(7.6), and

$$\begin{cases} \mathrm{F}_i(X_i(t)) = \begin{bmatrix} s_x(t) + \frac{u_p}{u_r}\cos(\theta(t) + T_0 u_r/2) \\ s_y(t) + \frac{u_p}{u_r}\sin(\theta(t) + T_0 u_r/2) \\ \theta(t) + T_0 u_r \\ \mathrm{g}_i(\mathrm{x}(t)) \end{bmatrix} + \begin{bmatrix} 0 & 0 \\ 0 & 1 - \gamma_i(t) \end{bmatrix} X_i(t), \\ \mathrm{G}_i(X_i(t)) = \gamma_i(t)\mathrm{g}_i(\mathrm{x}(t)) + \begin{bmatrix} 0 & 1 - \gamma_i(t) \end{bmatrix} X_i(t) + \gamma_i(t) D_i v_i(t). \end{cases} \tag{7.35}$$

Similarly, the linearized matrices $A_{F_i}(t)$ and $C_{G_i}(t)(i = 1, 2)$ of the nonlinear functions $\mathrm{F}_i(X_i(t))$ and $\mathrm{G}_i(X_i(t))$ can be calculated by

$$\begin{cases} A_{F_j}(t) = \begin{bmatrix} A_{f_j}(t) & \mathbf{0} \\ \gamma_i(t) C_{g_i}(t) & (1 - \gamma_i(t))\mathbf{I}_4 \end{bmatrix}_{\mathrm{x}(t)=\mathrm{x}^*}, \\ C_{G_i}(t) = \begin{bmatrix} \gamma_i(t) C_{g_i}(t) & (1 - \gamma_i(t))\mathbf{I}_4 \end{bmatrix}_{\mathrm{x}(t)=\mathrm{x}^*}. \end{cases} \tag{7.36}$$

Moreover, some relevant parameters are set as $T_0 = 1$, $u_p = 0.075$, $u_r = 2.5$, $D_1 = \mathrm{diag}\{0.6, 0.2, 0.4, 0.5\}$, $D_2 = \mathrm{diag}\{0.5, 0.6, 0.3, 0.8\}$, and $\gamma_i(t)$ is set to 0 with 50% probability. The system noises $w_p(t)$, $w_r(t)$, and $v_i(t)(i = 1, 2)$ are given as

$$\begin{cases} w_p(t) = 0.2\rho_p(t) - 0.1,\ w_r(t) = 0.3\rho_r(t) - 0.1, \\ v_1(t) = \mathrm{col}\left\{0.05\rho_{v_1}(t) - 0.02, 0.02\rho_{v_2}(t) - 0.01,\right. \\ \qquad \left. 0.03\rho_{v_3}(t) - 0.02, 0.05\rho_{v_4}(t) - 0.02\right\}, \\ v_2(t) = \mathrm{col}\left\{0.05\rho_{v_5}(t) - 0.01, 0.06\rho_{v_6}(t) - 0.02\right\}, \end{cases} \tag{7.37}$$

where $\rho_p(t)(\in [0, 1])$, $\rho_r(t)(\in [0, 1])$, and $\rho_{v_l}(t)(\in [0, 1])(l = 1, 2, 3, 4, 5, 6)$ are uniformly distributed random variables.

To show the effectiveness and advantages of the proposed secure fusion method summarized in Algorithm 7.1, the local tracking trajectories and the fusion trajectory are plotted in Fig. 7.2. It can be seen from this figure that the local estimator and the distributed fusion estimator all have good tracking performance against the DoS attacks.

Moreover, the mean-square error (MSE) and the sum of mean-square error (SMSE) are introduced to evaluate the estimation accuracy of the designed estimators, which are defined by

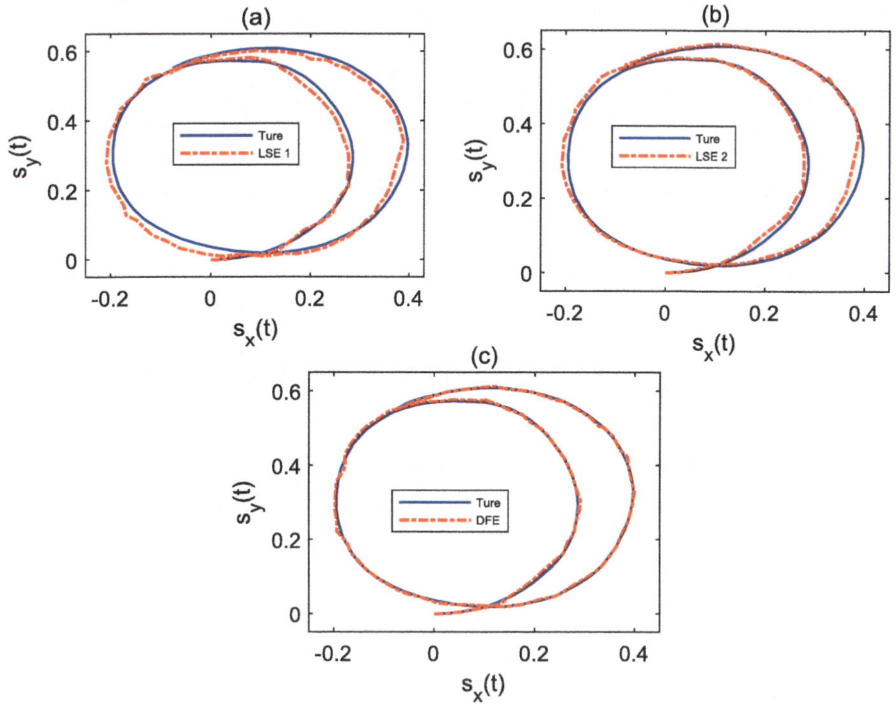

Fig. 7.2 The distributed secure fusion estimation structure for a class of nonlinear networked multi-sensor systems under DoS attacks

$$\mathrm{MSE}(t) = \frac{1}{M}\sum_{i=1}^{M}(x(t) - \hat{x}_i(t))^\top(x(t) - \hat{x}_i(t)), \tag{7.38}$$

$$\mathrm{SMSE}(t) = \frac{1}{M}\sum_{j=0}^{t}\sum_{i=1}^{M}(x(j) - \hat{x}_i(j))^\top(x(j) - \hat{x}_i(j)), \tag{7.39}$$

where $\hat{x}_i(t)$ is the local fusion estimation at time t, and M is the number of the Monte Carlo tests. In this case, 200 times Monte Carlo runs are performed to average the MSEs and SMSEs. Then, the MSEs of the LSEs and DFE are plotted in Fig. 7.3. It is seen from this figure that the MSE of DFE is smaller than the LSEs, which implies that the proposed secure fusion estimation algorithm indeed improves the tracking performance. Meanwhile, the SMSEs of the LSEs and DFE are depicted in Fig. 7.4, which further shows that the fusion estimation performance is much better than local estimation.

To further show the advantages of the secure fusion Algorithm 7.1, the performance comparison of the proposed method, EKF [9] method and UKF [10] method are presented. Specifically, Fig. 7.5 and Fig. 7.6 show the MSEs and SMSEs of these methods, respectively. It can be deduced from these figures that the developed esti-

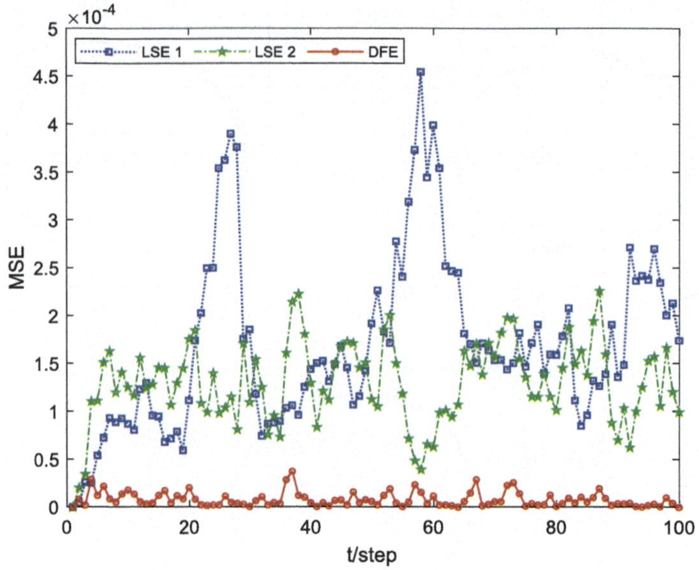

Fig. 7.3 MSE of the first LSE and DFE

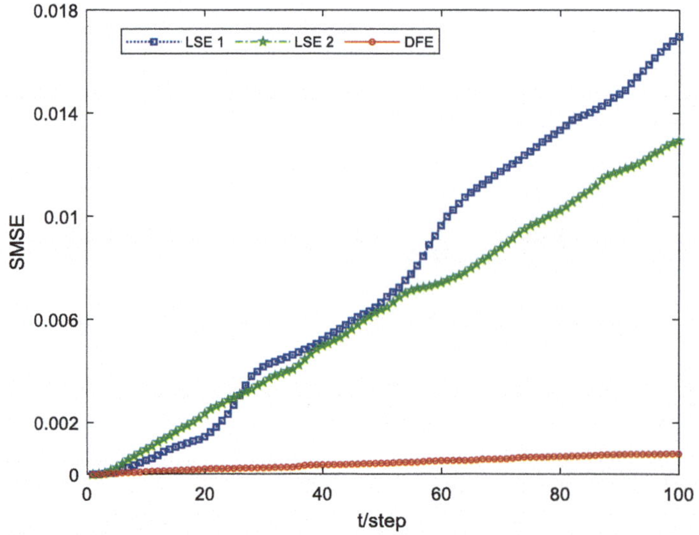

Fig. 7.4 SMSE of the first LSE and DFE

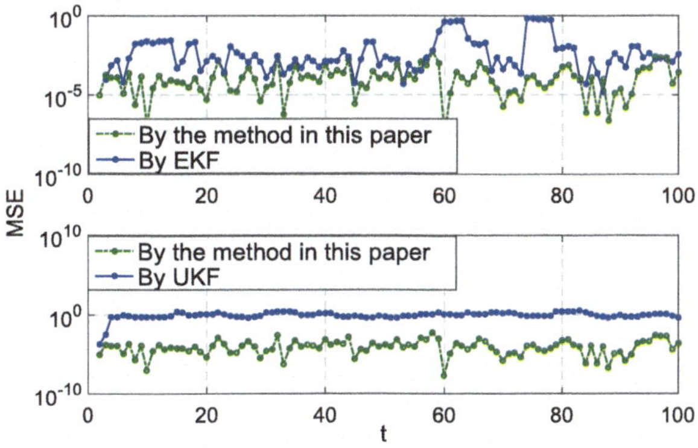

Fig. 7.5 The comparison of MSE by the proposed method, EKF and UKF

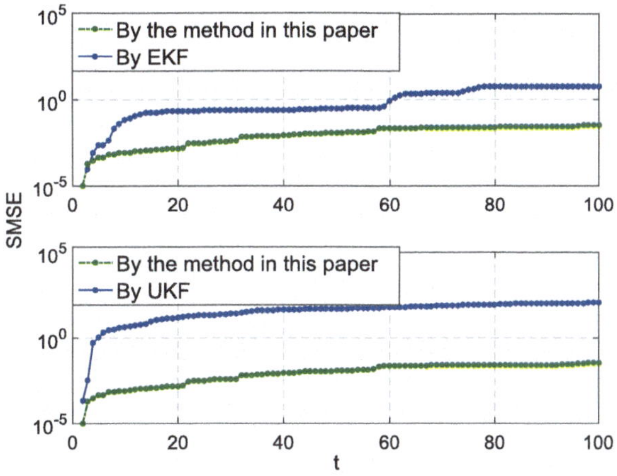

Fig. 7.6 The comparison of SMSE by the proposed method, EKF and UKF

mators in this chapter present a better fusion precision under the DoS attacks, which is mainly because the attack signal and noise statical information are difficult to be accurately calculated in practice, while the noise covariances need to be known in advance when implementing EKF and UKF methods. However, the system noises (7.37) are bounded without the statical information of noise, and the proposed method deals with this case based on the idea of bounded recursive optimization. Particularly, the proposed method can address the fusion estimation problem under Gaussian white noises with known covariances, which is more robust and has a wider range of applications.

7.5 Conclusion

In this chapter, a nonlinear secure fusion estimation algorithm was proposed for networked multi-sensor systems under DoS attacks, where the distributions of DoS attacks and noises were both not required to be known. Based on the idea of bounded recursive optimization, two convex optimization problems were established in terms of linear matrix inequality, which can be easily solved by the MATLAB LMI Toolbox to determine the local estimator gain and fusion criterion. Finally, an intelligent vehicle tracking system was employed to show the effectiveness and the advantages of the proposed methods. On the other hand, more practical and potential problems such as bandwidth constraints, energy constraints and time delay in networked systems will be investigated in the following chapters.

References

1. Carl G, Kesidis G, Brooks RR, Rai S (2006) Denial-of-service attack-detection techniques. IEEE Internet Comput 10(1):82–89
2. Chen B, Ho DWC, Hu G, Yu L (2018) Secure fusion estimation for bandwidth constrained cyber-physical systems under replay attacks. IEEE Trans Cybernet 48(6):1862–1876
3. Chen B, Ho DWC, Zhang W, Yu L (2019) Distributed dimensionality reduction fusion estimation for cyber-physical systems under DoS attacks. IEEE Trans Syst Man Cybernet Syst 49(2):455–468
4. Kordestani M, Chibakhsh A, Saif M (2019) A control oriented cyber-secure strategy based on multiple sensor fusion. In: 2019 IEEE International conference on systems, man and cybernetics (SMC), pp 1875–1881
5. Kordestani M, Dehghani M, Moshiri B, Saif M (2020) A new fusion estimation method for multi-rate multi-sensor systems with missing measurements. IEEE Access 8:47522–47532
6. Zhu D, Chen B, Yu L (2019) Distributed fusion estimation for linear time-varying systems under Dos attacks and bounded noises. In: 2019 Chinese control conference (CCC), pp 5480–5485
7. Chen B, Hu G, Ho Daniel W, C, Yu L (2019) A new approach to linear/nonlinear distributed fusion estimation problem. IEEE Trans Autom Control 64(3):1301–1308
8. Wang C, Yang M, Wang B, Xu H, Li H (2011) Improved intelligent vehicle localization using magnetic ruler. Int J Comput Intell Syst 4(3):394–401
9. Reif K, Gunther S, Yaz E, Unbehauen R (1999) Stochastic stability of the discrete-time extended Kalman filter. IEEE Trans Autom Control 44(4):714–728
10. Julier SJ, Uhlmann JK (1997) New extension of the Kalman filter to nonlinear systems. In: Kadar I (ed) Signal processing, sensor fusion, and target recognition VI. SPIE

Chapter 8
Active Secure Control Against DoS Attacks by Using Predictive Control Approach

8.1 Introduction

The security protection of CPSs involves various aspects, including privacy protection (discussed in Chaps. 2 and 3), attack detection (discussed in Chap. 4), security fusion assessment (discussed in Chaps. 5, 6, and 7), and the crucial consideration of secure control in the face of cyber attacks.

One prevalent form of cyber-attack is the Denial-of-Service (DoS) attack, which, notably, does not require much system information [1]. DoS attacks are particularly accessible in communication networks and are known to cause severe financial implications [2]. For instance, in the scenarios where a critical industrial control process is open-loop unstable, a DoS attack could lead to environmental damages. It is essential to note that if DoS attacks persist over an extended period, data packets may fail to update successfully in communication networks, hindering closed-loop feedback control. This introduces theoretical challenges to the analysis and synthesis of control systems.

This chapter will focus on investigating how to secure CPSs under DoS attacks. It is worth highlighting that existing approaches often assume DoS attacks occur only in the controller-to-actuator (C-A) or sensor-to-controller (S-C) channels [3–5], or simultaneously in both channels [6–8]. However, there is a more general scenario where DoS attacks may be asynchronous in two channels. In cases of asynchronous DoS attacks, even a small attack frequency can lead to prolonged failures in control signal updates compared with synchronous DoS attacks, making them more challenging to address [9].

In this chapter, an active secure control approach is developed for CPSs under asynchronous DoS attacks, where DoS attacks occur in both the S-C and C-A channels. Given the cost constraints of attacks, considering a limited number of maximum continuous DoS attacks in both channels is reasonable [10]. To defend against two-channel DoS attacks, an active secure control strategy is designed, making the optimal use of unattacked intervals to ensure timely control input updates. Additionally, a secure controller, encompassing both current and future control inputs, is

B. Chen et al., *Secure Fusion Estimation and Control for Cyber-Physical Systems Under Malicious Attacks*, https://doi.org/10.1007/978-981-97-9396-9_8

devised. Under the active secure control strategy and the secure controller, the CPS under two-channel DoS attacks is proven to be asymptotically stable without compromising control performance. Finally, simulations and experiments are provided to demonstrate the effectiveness of the proposed active secure control approach.

8.2 Problem Statement

Consider the CPS structure shown in Fig. 8.1. The physical process and system output are described by the following state-space model:

$$\begin{cases} x(k+1) = Ax(k) + Bu(k), \\ y(k) = Cx(k), \end{cases} \tag{8.1}$$

where $x \in \Re^n$ is the system state, $u \in \Re^m$ is the control input, and $y \in \Re^p$ is the system output. $A \in \Re^{n\times n}$, $B \in \Re^{n\times m}$ and $C \in \Re^{p\times n}$ are constant matrices.

This system model, represented in (8.1), finds widespread use in describing the dynamics of various systems, including power systems, smart grid infrastructures, and building automation systems. However, practical systems often experience actuator saturation issues due to physical limitations. Consequently, this chapter incorporates the consideration that the control input u in the system (8.1) is subject to

$$-\bar{u}_i \leq u_i \leq \bar{u}_i, i = 1, 2, \ldots, m, \tag{8.2}$$

where u_i is the ith element of control input u, and $\bar{u}_i$ is a known constant.

As shown in Fig. 8.1, the physical process is under network control, and both the S-C channel and the C-A channel in CPSs are attacked by the DoS attacks. Specifically, this chapter considers a more general scenario where the DoS attacks in the S-C channel and the C-A channel are independent. In other words, DoS attacks can occur in these two channels simultaneously or separately, presenting a more

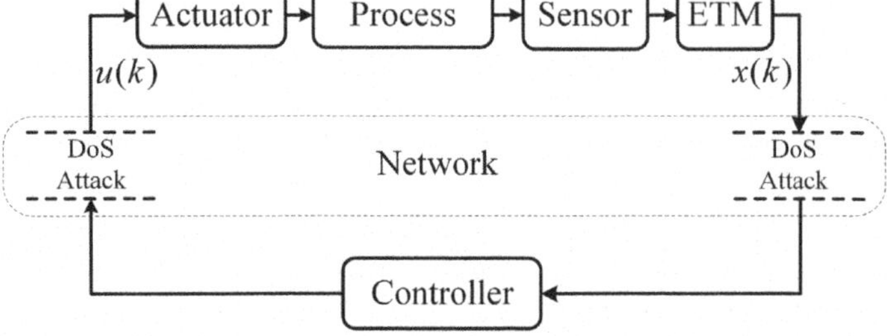

Fig. 8.1 The schematic diagram of the CPS under the two-channel DoS attacks. ETM is short for event-triggered mechanism

challenging scenario to address (see Remark 8.1). It is crucial to note that malicious attackers typically face cost constraints (e.g., limited energy) [10], making it difficult for them to sustain attacks continuously. Additionally, attackers may actively avoid continuous attacks to evade detection. Therefore, it is reasonable to assume that the number of maximum continuous DoS attacks in both the S-C channel and the C-A channel are bounded, denoted by N_{sc} and N_{ca} respectively.

Consequently, the objectives to be addressed in this chapter are outlined as follows.

- The first objective is to formulate an active secure control strategy that optimally utilizes the unattacked intervals and ensures timely updates of control inputs during each period under the influence of two-channel DoS attacks.
- The second objective is to design a secure controller that incorporates both current and future control inputs. This design should guarantee the asymptotic stability of the CPS represented by (8.1) under two-channel DoS attacks without compromising the control performance. Furthermore, the control inputs must satisfy the constraint given by (8.2).

Remark 8.1 In contrast to the scenarios typically addressed by the existing approaches, where DoS attacks occur either exclusively in the S-C channel or the C-A channel, or synchronously in both channels, the scenario considered in this chapter introduces increased complexity. Specifically, it is more challenging to successfully update control signals under the same attack frequency in both channels if the DoS attacks in the two channels are independent and asynchronous. For example, the control signals may not always be successfully updated if the attack frequency is 50% in each channel and the DoS attacks in the two channels are asynchronous. As a result, it becomes problematic to achieve feedback control over such networks. On the other hand, if the DoS attacks in the two channels are synchronous, 50% of the control signals can still be updated successfully. To address this difficulty, Sect. 8.3 introduces an event-based transmission policy. This policy aims to maximize unattacked intervals, ensuring successful update of measurement and control signals. This policy is critical to mitigate asynchronous DoS attack challenges in both channels.

8.3 Active Security Control Approach

8.3.1 Event-Based Security Transmission Policy

This subsection introduces an event-based secure transmission policy to actively defend against two-channel DoS attacks. The following event-triggered condition is considered first:

$$k_t = \begin{cases} \min\{k \mid \|x(k) - x(k_d)\| > \varepsilon\|x(k)\|\}, & k < k_d + H, \\ k_d + H, & k = k_d + H, \end{cases} \tag{8.3}$$

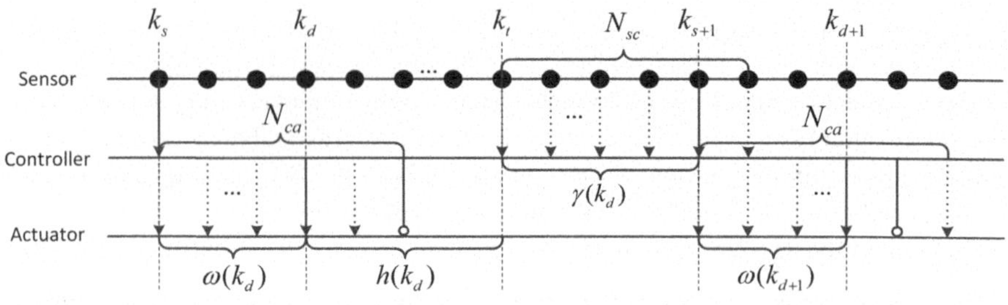

Fig. 8.2 The timing diagram of the CPSs with the active secure control strategy. "→" indicates that the data is sent successfully, "-->" indicates that the data fails to be sent, and "—o" represents the active packet loss

where $0 < \varepsilon \leq 1$. As shown in Fig. 8.2, k_d is the time when the control signal is successfully updated, k_t is the event-triggered time, and H is a positive integer.

In the scenario where the maximum number of continuous DoS attacks in the S-C channel is N_{sc} and C-A channel is N_{ca}, the following considerations apply:

(1) When the sensor sends measurement signals continuously $N_{sc} + 1$ times from the event-triggered time k_t to $k_t + N_{sc}$, it ensures that the controller successfully receives the measurement signal at least once. Let k_{s+1} denote the time when the controller successfully receives the measurement signal within the time interval k_t to $k_t + N_{sc}$.
(2) Similarly, from time k_s to $k_s + N_{ca}$, the controller sends control inputs continuously $N_{ca} + 1$ times, ensuring that the actuator successfully receives the control signal at least once.

In this case, we have $k_{d+1} \in \{k_d + 1, k_d + 2, \ldots, k_d + N\}$, where $N = H + N_{sc} + N_{ca}$. Let $h(k_d) = k_t - k_d$, $\omega(k_d) = k_d - k_s$ and $\gamma(k_d) = k_{s+1} - k_t$ and $N(k_d) = h(k_d) + \gamma(k_d) + \omega(k_{d+1})$, then it can be obtained that $h(k_d) \in \{1, 2, \ldots, H\}$, $\omega(k_d) \in \{0, 1, \ldots, N_{ca}\}$, $\gamma(k_d) \in \{0, 1, \ldots, N_{sc}\}$ and $N(k_d) \in \{1, 2, \ldots, N\}$. Specifically, when $k_{d+1} = k_d + N$, as shown in Fig. 8.2, the actuator does not receive control inputs for $N - 1$ periods due to the two-channel DoS attacks. This is the worst case that $h(k_d) = H$, $\gamma(k_d) = N_{sc}$ and $\omega(k_{d+1}) = N_{ca}$.

To sum up, the *event-based secure transmission policy* can be summarized as follows:

- At time k_d, the sensor executes event-triggered condition (8.3) to determine the event-triggered time k_t.
- From the time k_t to $k_t + N_{sc}$, the sensor sends the measurement signal to the controller $N_{sc} + 1$ times continuously.
- When the controller receive the measurement signal at time k_{s+1}, from time k_{s+1} to $k_{s+1} + N_{ca}$, the controller calculates the control signals and sends it to the actuator.

The introduced event-based secure transmission policy plays a crucial role in ensuring the successful updating of measurement signals and control signals. This is of paramount importance for achieving closed-loop feedback control under the influence of two-channel DoS attacks.

8.3.2 Security Controller Structure

In many existing methods, when communication channels are blocked by DoS attacks, the actuator typically employs a "sample-and-hold control" or "zero-input control" strategy. However, these strategies lead to a decrease in control performance due to the inability to update the control signal in a timely manner. To address this issue, this chapter proposes a *secure controller* that encompasses both current and future control inputs, actively defending against DoS attacks.

As shown in Fig. 8.2, the controller calculates N_C control inputs at time k_d (i.e., $u(k_d), \ldots, u(k_d+i), \ldots, u(k_d+N_C-1)$), and then sends the N_C control inputs to the actuator by using one data packet. Upon receiving the data packet, the actuator sequentially executes the control inputs $u(k_d), \ldots, u(k_d+i), \ldots, u(k_d+N_C-1)$ from k_d to k_{d+1}. Under the event-based secure transmission policy, the actuator may not receive control inputs for up to $N-1$ periods. In such cases, when the number of control inputs N_C is not less than N (i.e., $N_C \geq N$), appropriate control inputs can be updated in each period, even in the presence of two-channel DoS attacks in CPSs. This active approach is referred to as an *active secure control approach*. Specifically, a state feedback controller is considered as

$$u(k_d+i) = Kx(k_d+i\,|k_d) = K(A+BK)^i x(k_d), i = 0, 1, \ldots, N_C-1, \quad (8.4)$$

where $x(k_d) \triangleq x(k_d\,|k_d)$, $u(k_d+i)$ is the control input at time k_d+i, and $x(k_d+i\,|k_d)$ is the state at time k_d+i predicted at time k_d. Notice that, as shown in Fig. 8.2, the controller may not be able to obtain the state $x(k_d)$ at time k_d, but it can obtain the state $x(k_s)$ at time k_s. Hence, a predictive state $x(k_d\,|k_d-\omega(k_d))$ is given by:

$$x(k_d\,|k_d-\omega(k_d)) = x(k_s+\omega(k_d)\,|k_s) = (A+BK)^{\omega(k_d)} x(k_s). \quad (8.5)$$

Furthermore, the controller (8.4) is rewritten as

$$\begin{aligned} u_{\omega(k_d)}(k_d+i) &= Kx(k_d+i\,|k_d-\omega(k_d)) \\ &= K(A+BK)^i x(k_d\,|k_d-\omega(k_d)), \end{aligned} \quad (8.6)$$

where $i = 0, 1, \ldots, N_C-1$. Let $U_{\omega(k_d)}(k_d)$ stand for control sequence

$$\{u_{\omega(k_d)}(k_d), \ldots, u_{\omega(k_d)}(k_d+i), \ldots, u_{\omega(k_d)}(k_d+N_C-1)\}.$$

Then, the controller calculates $U_{\omega(k_d)}(k_d)$ at time k_d, and then sends it to the actuator.

Combined with the event-based secure transmission policy, the *active secure control strategy* is summarized in Algorithm 8.1.

Algorithm 8.1 Active Security Control Strategy

- Step 1: When the actuator receives the control sequence $U_{\omega(k_d)}(k_d)$ at time k_d, the actuator executes the control inputs $u_{\omega(k_d)}(k_d), \ldots, u_{\omega(k_d)}(k_d + i), \ldots, u_{\omega(k_d)}(k_d + N_C - 1)$ successively, while the sensor executes event-triggered condition (8.3) to determine the event-triggered time k_t;
- Step 2: At time k_t, the sensor sends the state vector to the controller $N_{sc} + 1$ times continuously from k_t to $k_t + N_{sc}$;
- Step 3: After the controller receives the state $x(k_{s+1})$ at time k_{s+1}, the controller constructs the control sequence $U_{\omega(k_{d+1})}(k_{d+1})$ based on the $x(k_{s+1} + \omega(k_{d+1})|k_{s+1})$ and then sends it to the actuator $N_{ca} + 1$ times continuously from k_{s+1} to $k_{s+1} + N_{ca}$. When the actuator receives the control sequence $U_{\omega(k_{d+1})}(k_{d+1})$, let $k_d \rightarrow k_{d+1}$ and $k_s \rightarrow k_{s+1}$. Return to the Step 1.

Remark 8.2 In most communication networks, data is transmitted in "packets", and whether sending one bit or hundreds of bits within its payload, it consumes the same amount of network resources [11]. Therefore, as long as all the control signals in $U_{\omega(k_d)}(k_d)$ can be packed into one data packet, the consumption of network resources will not increase, which implies that the N_C should be chosen appropriately to satisfy the following inequality:

$$N_C \leq \frac{B_p}{B_c} - 1, \tag{8.7}$$

where B_p is the bits of effective load of the data packet and B_c is the bits of a single control signal. It should be pointed out that the inequality (8.7) is easy to be satisfied in practice. For example, in the User Datagram Protocol (UDP), $B_p = 1472$ bits. For $B_c = 8$ bits, $N_C \leq 183$, and it is sufficiently large for most applications.

Remark 8.3 As shown in Fig. 8.2, from time k_s to $k_s + N_{ca}$, the controller calculates the control sequence $U_{\omega(k_d)}(k_d)$ based on the $x(k_s)$ and sends it to the actuator. Then the actuator may receive multiple control sequences from time k_s to $k_s + N_{ca}$. For this case, we design a simple execution rule. That is, the first control sequence $U_{\omega(k_d)}(k_d)$ received by the actuator will be implemented, and the subsequent control sequences will be actively discarded. The reason is that, from time k_s to $k_s + N_{ca}$, all the control sequences are calculated based on the $x(k_s)$, and the subsequent control sequences does not contain new feedback information.

Remark 8.4 In [3–5, 9], when the measurement signal is sent to the controller in the S-C channel, the controller needs to send an acknowledge (ACK) signal back to the sensor via network to confirm the success of the transmission. This is the same as in the C-A channel. However, the ACK signal may also be attacked, which leads to the sensor or controller to be unknown about whether the data is transmitted successfully. Fortunately, under the proposed active secure control strategy in Algorithm 8.1, the ACK signal is not required.

Theorem 8.1 *If the number of control inputs N_C in $U_{\omega(k_d)}(k_d)$ is not less than N, i.e., $N_C \geq N$, the active secure control strategy in Algorithm 8.1 can guarantee that the control inputs are updated timely in each period under the two-channel DoS attacks, where $N = H + N_{sc} + N_{ca}$.*

Proof For the active secure control strategy in Algorithm 8.1, at time k_d, the actuator receives the control sequence $U_{\omega(k_d)}(k_d)$ that contains both the current and future control inputs. Then, due to the two-channel DoS attacks, the actuator may not receive control sequence for up to $N - 1$ periods. If $N_C \geq N$, according to the active secure control strategy, the control inputs $u_{\omega(k_d)}(k_d)$, $u_{\omega(k_d)}(k_d + 1), \ldots, u_{\omega(k_d)}(k_d + j - 1)$ will be implemented successively in the case of $k_{d+1} = k_d + j$, $(j \in \{1, 2, \ldots, N\})$. Then, at time $k_{d+1} = k_d + j$, a new control sequence will be received by actuator, and the new control inputs will be implemented. Therefore, it can be obtained that $k_{d+1} \in \{k_d + 1, \ldots, k_d + j, \ldots, k_d + N\}$. Specifically, the system dynamics can be described by the following N cases:

Case 1: when $k_{d+1} = k_d + 1$,

$$\begin{aligned} x(k_{d+1}) &= x(k_d + 1) = Ax(k_d) + Bu_{\omega(k_d)}(k_d) \\ &= (A + BK)x(k_d) - BK(x(k_d) - x(k_d | k_d - \omega(k_d))) \\ &= (A + BK)x(k_d), \end{aligned} \tag{8.8}$$

where

$$\begin{aligned} & x(k_d) - x(k_d | k_d - \omega(k_d)) \\ &= Ax(k_d - 1) - Ax(k_d - 1 | k_d - \omega(k_d)) \\ &\cdots \\ &= A^{\omega(k_d)}(x(k_d - \omega(k_d)) - x(k_d - \omega(k_d) | k_d - \omega(k_d))) = \mathbf{0}. \end{aligned} \tag{8.9}$$

Case 2: when $k_{d+1} = k_d + 2$,

$$\begin{aligned} x(k_{d+1}) &= x(k_d + 2) \\ &= Ax(k_d + 1) + Bu_{\omega(k_d)}(k_d + 1) \\ &= Ax(k_d + 1) + BKx(k_d + 1) - BKx(k_d + 1) + BKx(k_d + 1 | k_d - \omega(k_d)) \\ &= (A + BK)x(k_d + 1) - BK(x(k_d + 1) - x(k_d + 1 | k_d - \omega(k_d))) \\ &= (A + BK)^2 x(k_d). \end{aligned} \tag{8.10}$$

$$\vdots$$

Case j: when $k_{d+1} = k_d + j$,

$$x(k_{d+1}) = x(k_d + j) = (A + BK)x(k_d + j - 1) = (A + BK)^j x(k_d). \quad (8.11)$$

$$\vdots$$

Case N: when $k_{d+1} = k_d + N$,

$$x(k_{d+1}) = x(k_d + N) = (A + BK)x(k_d + N - 1) = (A + BK)^N x(k_d). \quad (8.12)$$

Therefore, the closed-loop system model is derived as follows:

$$x(k_{d+1}) = x(k_d + N(k_d)) = (A + BK)^{N(k_d)} x(k_d), \quad (8.13)$$

where $N(k_d) \in \{1, 2, \ldots, N\}$. It should be pointed out that the closed-loop system (8.13) is equivalent to the case without DoS attacks. Based on the analysis above, when $N_C \geq N$, there are appropriate control inputs to be updated in each period under the two-channel DoS attacks by implementing Algorithm 8.1.

Remark 8.5 Under the active secure control strategy, the "worst case" of $k_{d+1} = k_d + N$ is that the actuator does not receive control sequence for $N - 1$ periods. Benefiting from the structure of secure controller, the "worst case" of $k_{d+1} = k_d + N$ can be addressed. It implies that the CPSs can tolerate continuous DoS attacks in S-C channel between k_s and $k_s + N$ or C-A channel between k_d and $k_d + N$, where $N = H + N_{sc} + N_{ca}$. That is, the number of maximum continuous DoS attacks is $N - 1$. In (8.3), when $H = \mathbf{0}$, the event-triggered mechanism (ETM) degenerates into time-triggered mechanism. Then, the number of maximum continuous DoS attacks is reduced to $N = N_{sc} + N_{ca} - 1$. Therefore, the ETM can increase the number of maximum admissible continuous DoS attacks of CPSs.

8.4 Stability Analysis and Controller Design

Though the proposed active secure control strategy can ensure that the control inputs are updated timely in each period, the design procedure of control sequence $U_{\omega(k_d)}(k_d)$ needs to satisfy the desired asymptotically stability criteria, and the control inputs must satisfy the constraint (8.2). Then, the predictive control theory [12] is employed to design the stable controller in this chapter. For the CPSs (8.13), the following objective function at time k_d is considered:

$$J(k_d) = \sum_{i=0}^{\infty} [\|x(k_d + i\,|k_d)\|_R^2 + \|u(k_d + i\,|k_d)\|_S^2], \quad (8.14)$$

where $R > \mathbf{0}$ and $S > \mathbf{0}$ are symmetric weighting matrices. Next, the control invariant ellipsoid of system (8.1) under control law (8.6) is defined as [12]

$$E_x = \left\{x \in \Re^n \left| x^{\mathrm{T}} Q_x^{-1} x \leq 1 \right.\right\}, Q_x^{-1} \geq \mathbf{0}. \tag{8.15}$$

Then, if

$$(A + BK)^{\mathrm{T}} Q_x^{-1} (A + BK) \leq Q_x^{-1}, \tag{8.16}$$

it can be easily obtained that

$$x(k_d) \in E_x \Rightarrow x(k_d + 1) \in E_x. \tag{8.17}$$

Furthermore, it follows from (8.16) and (8.17) that

$$x(k_d) \in E_x \Rightarrow x(k_d + N(k_d)) = x(k_{d+1}) \in E_x. \tag{8.18}$$

To obtain an upper bound of $J(k_d)$, the Lyapunov function $V(x) = x^{\mathrm{T}} P_x x$ is considered. Then, if

$$(A + BK)^{\mathrm{T}} P_x (A + BK) - P_x \leq -(R + K^{\mathrm{T}} SK), \tag{8.19}$$

it can ensure that

$$\begin{aligned}
&V(x(k_d + i + 1 \,|k_d)) - V(x(k_d + i \,|k_d)) \\
&\leq -x^{\mathrm{T}}(k_d + i \,|k_d) R x(k_d + i \,|k_d) \\
&- u^{\mathrm{T}}(k_d + i \,|k_d) S u(k_d + i \,|k_d), i = 0, 1, \ldots, \infty.
\end{aligned} \tag{8.20}$$

It can be concluded from (8.20) that closed-loop system (8.13) is asymptotically stable. Then $x(\infty \,|k_d) = 0$ and $V(x(\infty \,|k_d)) = 0$. Summing up the inequation (8.20) from $i = 0$ and $i = \infty$, one has

$$-V(x(k_d)) \leq -J(k_d) \Rightarrow J(k_d) \leq V(x(k_d)). \tag{8.21}$$

Defining $P_x = Q_x^{-1}$ and $K = Y Q_x^{-1}$, we can see that (8.16) is derived from (8.19). By using the Schur complement lemma [13], the inequality (8.19) can be converted to

$$\begin{bmatrix} Q_x & Q_x A^{\mathrm{T}} + Y^{\mathrm{T}} B^{\mathrm{T}} & Q_x & Y^{\mathrm{T}} \\ AQ_x + BY & Q_x & \mathbf{0} & \mathbf{0} \\ Q_x & \mathbf{0} & R^{-1} & \mathbf{0} \\ Y & \mathbf{0} & \mathbf{0} & S^{-1} \end{bmatrix} \geq \mathbf{0}. \tag{8.22}$$

The input constraints in (8.2) can be rewritten as

$$\left|u_j\right| = \left|K_j x\right| \le \bar{u}_j,\ j = 1, 2, \ldots, m, \tag{8.23}$$

where K_j denotes the jth row of the matrix K. For $x \in E_x$, (8.23) implies

$$\begin{aligned}\left|K_j x\right|^2 &= |K_j Q_x^{1/2} Q_x^{-1/2} x|^2 \le \|K_j Q_x^{1/2}\|^2 \|Q_x^{-1/2} x\|^2 \\ &= K_j Q_x K_j^{\mathrm{T}} x^{\mathrm{T}} Q_x^{-1} x \le K_j Q_x K_j^{\mathrm{T}}.\end{aligned} \tag{8.24}$$

Let

$$K_j Q_x K_j^{\mathrm{T}} \le \bar{u}_j^2, \quad j = 1, 2, \ldots, m. \tag{8.25}$$

Thus, the condition (8.2) holds under (8.24) and (8.25). Moreover, (8.25) is equivalent to

$$\begin{bmatrix} Z & Y \\ Y^{\mathrm{T}} & Q_x \end{bmatrix} \ge 0, \quad Z_{jj} \le \bar{u}_j^2, \quad j = 1, 2, \ldots, m, \tag{8.26}$$

where Z is a symmetric matrix.

Inspired by the idea of predictive control in [12], an additional term $c(k)$ is introduced in (8.6) to reduce online computation and improve control performance, i.e.,

$$u_{\omega(k_d)}(k_d + i) = Kx(k_d + i\ |k_d - \omega(k_d)) + c(k_d + i), \tag{8.27}$$

where $c(k_d + i)$, $i = 0, \ldots, n_c - 1$, represents the degree of design freedom, and $c(k_d + n_c + i) = 0$, $i \ge 0$. Then, the closed-loop system of (8.1) is written as

$$x(k_d + 1) = (A + BK)x(k_d) + Bc(k_d). \tag{8.28}$$

Defining

$$\begin{aligned} f(k_d) &= \mathrm{col}\{c(k_d), c(k_d + 1), \ldots, c(k_d + n_c - 1)\}, \\ z(k_d) &= \mathrm{col}\{x(k_d), f(k_d)\}, \end{aligned}$$

then (8.28) is described by

$$\begin{aligned} z(k_d + 1) &= \Psi z(k_d), \quad z \in \Re^{n+mn_c}, \\ \Psi &= \begin{bmatrix} A + BK & BD \\ \mathbf{0} & M \end{bmatrix}, M = \begin{bmatrix} \mathbf{0}_{(n_c-1)m\times m} & \mathbf{I}_{(n_c-1)m} \\ \mathbf{0}_{m\times m} & \mathbf{0}_{m\times(n_c-1)m} \end{bmatrix}, \\ D &= \begin{bmatrix} \mathbf{I}_m & \mathbf{0}_{m\times(n_c-1)m} \end{bmatrix}. \end{aligned} \tag{8.29}$$

For the system (8.29), if condition

$$\Psi^{\mathrm{T}} Q_z^{-1} \Psi \le Q_z^{-1} \tag{8.30}$$

holds, the state $z(k_d)$ shall fall into the invariant ellipsoid

$$E_z = \left\{ z \in \Re^{n+mn_c} \left| z^{\mathrm{T}} Q_z^{-1} z \leq 1 \right. \right\}, Q_z^{-1} \geq 0. \tag{8.31}$$

Then, it is obtained from (8.30) and (8.31) that

$$z(k_d) \in E_z \Rightarrow z(k_d + N(k_d)) = z(k_{d+1}) \in E_z. \tag{8.32}$$

By choosing the Lyapunov function $V(z) = z^{\mathrm{T}} P_z z$ for (8.29), if

$$\Psi^{\mathrm{T}} P_z \Psi - P_z \leq -(T^{\mathrm{T}} R T + T^{\mathrm{T}} K^{\mathrm{T}} S K T), \tag{8.33}$$

where $T = \left[\mathbf{I}_n \ \mathbf{0}_{n \times mn_c} \right]$, $x = Tz$, one has

$$\begin{aligned} & V(z(k_d + i + 1 \,|k_d)) - V(z(k_d + i \,|k_d)) \\ & \leq -x^{\mathrm{T}}(k_d + i \,|k_d) R x(k_d + i \,|k_d) \\ & - u^{\mathrm{T}}(k_d + i \,|k_d) S u(k_d + i \,|k_d). \end{aligned} \tag{8.34}$$

Similar to (8.21), it can be obtained from (8.33) that

$$-V(z(k_d)) \leq -J(k_d) \Rightarrow J(k_d) \leq V(z(k_d)). \tag{8.35}$$

Defining $P_z = Q_z^{-1}$, we can see that (8.30) is derived from (8.33). Applying the Schur complement lemma, the inequality (8.33) can be converted to

$$\begin{bmatrix} Q_z & Q_z \Psi^{\mathrm{T}} & Q_z T^{\mathrm{T}} & Q_z T^{\mathrm{T}} K^{\mathrm{T}} \\ \Psi Q_z & Q_z & \mathbf{0} & \mathbf{0} \\ T Q_z & \mathbf{0} & R^{-1} & \mathbf{0} \\ K T Q_z & \mathbf{0} & \mathbf{0} & S^{-1} \end{bmatrix} \geq \mathbf{0}. \tag{8.36}$$

Then, the inequation (8.25) can be rewritten as

$$\left[K_j \ e_j \ \mathbf{0} \ldots \mathbf{0} \right] Q_z \left[K_j \ e_j \ \mathbf{0} \ldots \mathbf{0} \right]^{\mathrm{T}} \leq \bar{u}_j^2, \qquad j = 1, 2, \ldots, m, \tag{8.37}$$

where e_j denotes the jth row of the identity matrix.

On the other hand, applying a congruence transformation to (8.31), one has

$$\begin{bmatrix} T & \mathbf{0} \\ \mathbf{0} & \mathbf{I} \end{bmatrix} \begin{bmatrix} Q_z & z \\ z^{\mathrm{T}} & 1 \end{bmatrix} \begin{bmatrix} T^{\mathrm{T}} & \mathbf{0} \\ \mathbf{0} & \mathbf{I} \end{bmatrix} \geq 0 \Rightarrow \begin{bmatrix} T Q_z T^{\mathrm{T}} & x \\ x^{\mathrm{T}} & 1 \end{bmatrix} \geq 0. \tag{8.40}$$

Define $E_{xz} = \left\{ x \in \Re^n \left| x^{\mathrm{T}} Q_{xz}^{-1} x \leq 1 \right. \right\}$, where $Q_{xz}^{-1} \geq 0$, and $Q_{xz} = T Q_z T^{\mathrm{T}}$. Then, due to the advantages of degrees of design freedom of f, a larger volume of invariant ellipsoid E_{xz} than E_x can be obtained by minimizing $\log\det(T Q_z T^{\mathrm{T}})^{-1}$ subject to LMIs (8.36) and (8.37) [13], which implies $E_x \subseteq E_{xz}$. Since LMIs (8.36) and

Algorithm 8.2 Controller Design Procedure

Offline design:

- Step 1: Solve the LMIs (8.22) and (8.26) offline, and obtain the feedback control law K;
- Step 2: For this control law K, solve the following optimization problem to maximize the volume of E_{xz}:

$$\begin{aligned} &\min_{Q_z} \log\det(TQ_zT^{\mathrm{T}})^{-1} \\ &s.t. \quad (8.36) \text{ and } (8.37). \end{aligned} \tag{8.38}$$

Online synthesis:

At time k_d, the controller solves the following online optimization problem to obtain the $f(k_d)$:

$$\begin{aligned} &\min_{f(k_d)} f^{\mathrm{T}}(k_d)f(k_d) \\ &\text{s.t.} \quad z^{\mathrm{T}}(k_d)Q_z^{-1}z(k_d) \le 1. \end{aligned} \tag{8.39}$$

Then, the controller calculates the control inputs (8.27) and further constructs the control sequence $U_{\omega(k_d)}(k_d)$.

(8.37) do not depend on the current state, the algorithm can be implemented offline. Now, the controller design procedure of "*Offline design* and *Online synthesis*" can be summarized as Algorithm 8.2.

Theorem 8.2 *By applying Algorithm 8.2 and the active secure control strategy in Algorithm 8.1, the closed-loop system* (8.13) *under the two-channel DoS attacks is asymptotically stable without losing the control performance, and the control inputs satisfy the constraints* (8.2).

Proof As the analysis above, in Algorithm 8.2, the LMIs (8.22), (8.26) and the optimization problem (8.38) as well as the online optimization problem (8.38) can ensure that the closed-loop system (8.13) is asymptotically stable, and the control inputs satisfy the constraint (8.2). As stated in Theorem 8.1, under the two-channel DoS attacks, the active secure control strategy can ensure that the control inputs are updated timely in each period, and the resulting closed-loop system (8.13) is equivalent to the case without DoS attacks. Therefore, the proposed active secure control approach can achieve the same control performance as the cases without DoS attacks.

Remark 8.6 The LMIs (8.22) and (8.26), as well as the optimization problem (8.38) can be solved by the LMI Toolbox in Matlab [13]. Meanwhile, the online optimization problem (8.39) can be solved by the search algorithm in [14], and it has been shown in [14] that this online search algorithm only requires short CPU time.

Remark 8.7 If the state vector x is not available, an observer located between the sensor and the ETM can be introduced as follows:

$$\hat{x}(k+1) = A\hat{x}(k) + Bu(k) + L(y(k) - C\hat{x}(k)), \tag{8.41}$$

where $\hat{x}(k)$ is the observed state at time k. L is the observer gain matrix guaranteeing that $A - LC$ is Schur-stable. Let the observation error be $e(k) = x(k) - \hat{x}(k)$. Then, there will be $\lim_{k\to\infty} e(k) = 0$. Furthermore, an observer-based active secure control approach can be easily obtained from the similar derivation in this chapter.

8.5 Simulation and Experiment

A networked inverted pendulum control system (see Fig. 8.3) is considered in this section to show the effectiveness of the proposed methods by the simulations and experiments, while the corresponding structure diagram can be represented by Fig. 8.4, where the data packets are transmitted over communication networks using the User Datagram Protocol (UDP). By using the Newton-Euler method, the discrete-time state-space model of inverted pendulum system with sampling period $T = 5$ ms is obtained as follows:

$$x(k + 1) = Ax(k) + Bu(k), \tag{8.42}$$

where $A = \begin{bmatrix} 1 & 0.0050 & 0 & 0 \\ 0 & 1 & 0 & 0 \\ 0 & 0 & 1.0004 & 0.0050 \\ 0 & 0 & 0.1470 & 1.0004 \end{bmatrix}$, $B = \begin{bmatrix} 0 \\ 0.005 \\ 0 \\ 0.015 \end{bmatrix}$, $x = \begin{bmatrix} p & \dot{p} & \phi & \dot{\phi} \end{bmatrix}^{\mathrm{T}}$, p is the position of the cart, $\dot{p}$ is the velocity of the cart, $\phi = \theta - \pi$ with θ being the angle between the pendulum and the vertical downward direction, and $\dot{\phi}$ is the angular velocity of the pendulum. The control input u in (8.42) is subjected to the following constraints:

$$-18m/s^2 \leq u \leq 18m/s^2. \tag{8.43}$$

That is, the acceleration of the cart moving to the left or right is no more than $18m/s^2$. Consider the case where the number of maximum continuous DoS attacks in S-C channel and C-A channel are $N_{sc} = 3$ and $N_{ca} = 3$, respectively.

8.5.1 Simulation

Let $R = \mathrm{diag}\{80, 3, 10, 1\}$ and $S = 1$. By implementing Algorithm 8.2, the following controller gain matrix is obtained:

$$K = [\, 11.606010.5145\ -54.7776 - 10.1485 \,]. \tag{8.44}$$

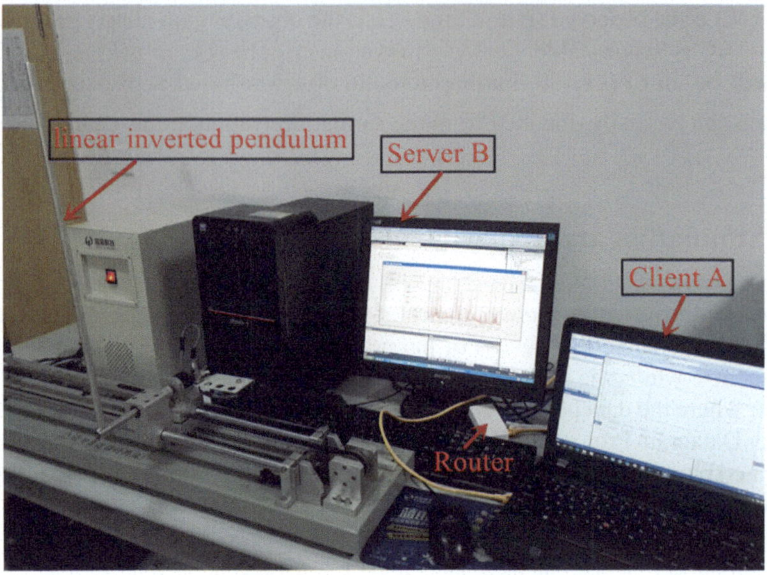

Fig. 8.3 The networked inverted pendulum control system platform

Fig. 8.4 The structure diagram of the networked inverted pendulum control system

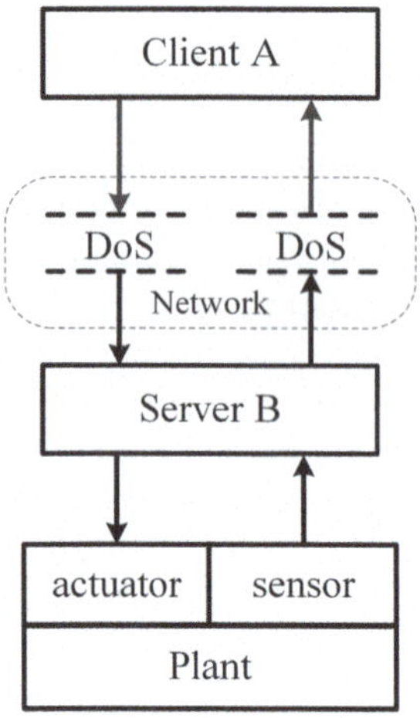

Considering $\varepsilon = 0.9$, $H = 3$, $N_{sc} = 3$ and $N_{ca} = 3$, it can be obtained that $N = 9$. Let $N_C = N = 9$ and $n_c = 9$. Then, applying Algorithm 8.2 and the active secure control strategy in Algorithm 8.1, the simulation results are shown in Figs. 8.5 and 8.6.

The system outputs and control inputs are plotted in Fig. 8.5. To clearly illustrate the implementation process of the active secure control strategy, the feedback control timing diagram of the first 50 periods under the two-channel DoS attacks is shown by Fig. 8.6. It is seen from Fig. 8.5 that i) The proposed active secure control approach

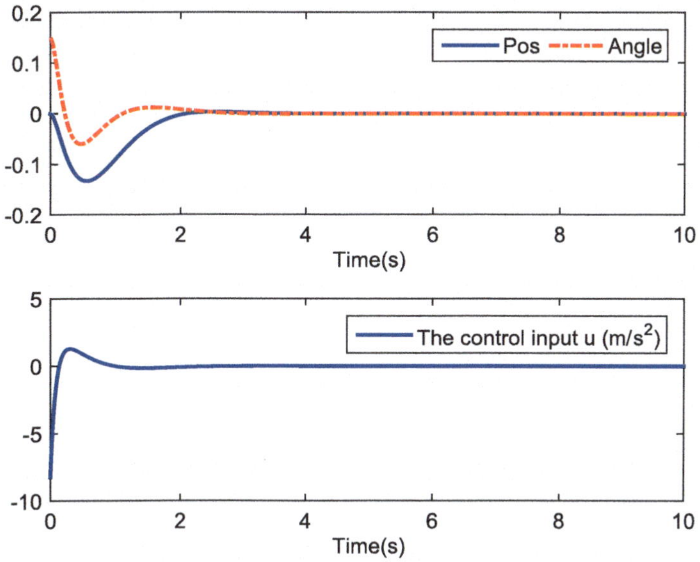

Fig. 8.5 The system outputs and control inputs in the simulation

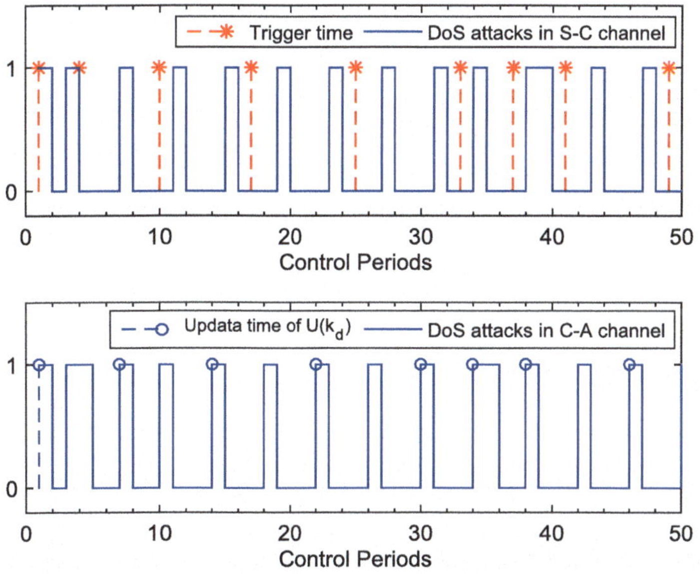

Fig. 8.6 The feedback control timing diagram under the two-channel DoS attacks in the simulation. The time of 0 to 1 and 1 represent no DoS attacks in the communication channel, and the time of 1 to 0 and 0 represent DoS attacks in the communication channel. "$*$" represents the event-trigger time, and "$\circ$" represents the update time of $U_{\omega(k_d)}(k_d)$

can guarantee the asymptotic stability of the networked inverted pendulum control system (8.42) without losing the control performance; ii) The control inputs satisfy the constraint (8.43). Besides, it is seen from Fig. 8.6 that the active secure control strategy can ensure that the control inputs are updated timely in each period. This accords with the results in Theorem 8.2, i.e., the active secure control approach proposed in this chapter can achieve the same control performance as the cases without DoS attacks.

8.5.2 *Experiment*

8.5.2.1 Case 1

By setting the same parameters and network environment as in the simulations and applying the active secure control approach proposed in this chapter, the experiment results are shown in Figs. 8.7 and 8.8. It can be seen from Fig. 8.7 that the networked inverted pendulum control system is stable, and the control inputs satisfy the constraint (8.43). Figure 8.8 shows the feedback control timing diagram of the first 50 periods under the two-channel DoS attacks in the experiment. It can be seen from this figure that the control sequence $U_{\omega(k_d)}(k_d)$ containing both the current and future control inputs is updated 10 times successfully during the first 50 periods. When the actuator receives the control sequence $U_{\omega(k_d)}(k_d)$, the control inputs $u_{\omega(k_d)}(k_d), u_{\omega(k_d)}(k_d+1), \ldots, u_{\omega(k_d)}(k_d+j-1)$ will be implemented successively in the case of $k_{d+1} = k_d + j$, $(j \in \{1, 2, \ldots, 9\})$. Therefore, Fig. 8.8 demonstrates that there must be appropriate control inputs to be updated in each period under the two-channel DoS attacks by implementing the active secure control strategy. Figures 8.7 and 8.8 show that the active secure control approach proposed in this chapter is also effective in the real system.

8.5.2.2 Comparative Experiment

In this subsection, we will give a comparative experiment to verify the advantages of the active secure control approach proposed in this chapter. Different from the active secure control strategy in Algorithm 8.1, in comparative experiments, a "sample-and-hold control" strategy is considered, where the controller only calculates a single control input $u_{\omega(k_d)}(k_d)$ and then sends it to the actuator. When the actuator receives the control input $u_{\omega(k_d)}(k_d)$ at time k_d, the actuator executes the control input $u_{\omega(k_d)}(k_d)$ until a new control input is received, i.e., $u(k) = u_{\omega(k_d)}(k_d)$, $k_d \le k < k_{d+1}$. Compared with Algorithm 8.1, the sample-and-hold control strategy in comparative experiments will reduce the control performance because the control input cannot be updated timely.

Considering the same parameters and network environment as in Case 1, the comparative experiment results are shown in Fig. 8.9, where the position and angle

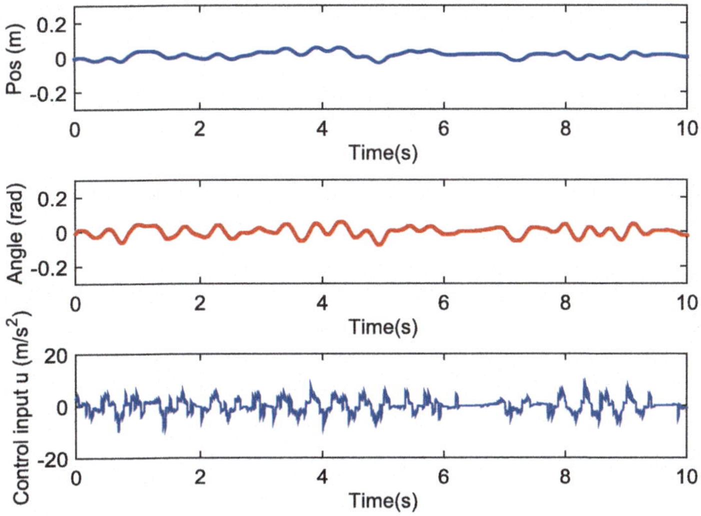

Fig. 8.7 The system outputs and control inputs in the experiment

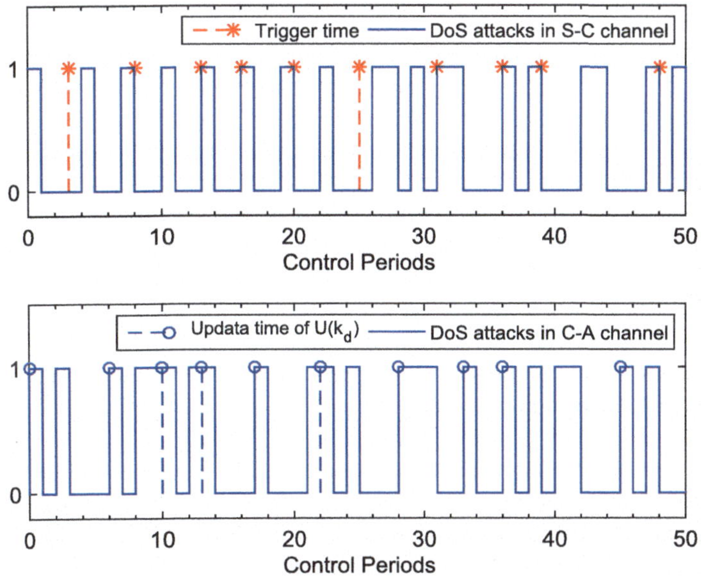

Fig. 8.8 The feedback control timing diagram under the two-channel DoS attacks in the experiment

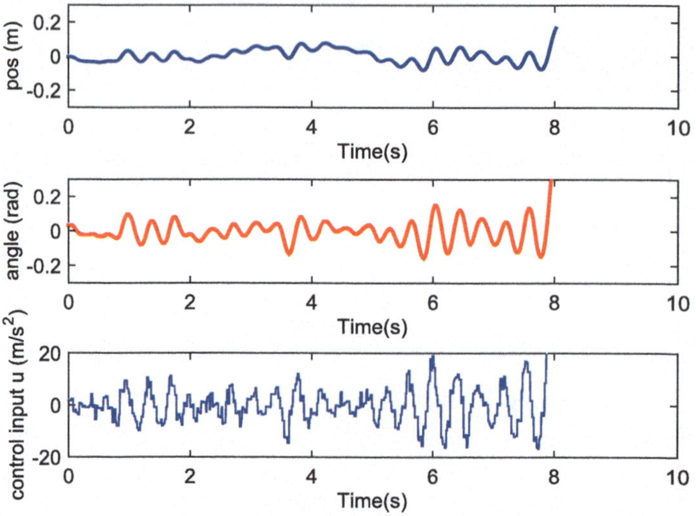

Fig. 8.9 The system outputs and control inputs in the comparative experiment

curves of the inverted pendulum fluctuate greatly, and the inverted pendulum becomes unstable around the 8th second after running for a short time. Compared with Fig. 8.7, the experiment results confirm that the active secure control approach proposed in this chapter is superior to the sample-and-hold control strategy considered in the comparative experiments.

8.6 Conclusion

An active secure control approach has been proposed for CPSs under the two-channel DoS attacks, where the number of maximum continuous DoS attacks in both the S-C and the C-A channels is considered to be bounded due to the cost constraints of attacks. To actively defend against the two-channel DoS attacks, the active secure control strategy in Algorithm 8.1 has been developed which can ensure that there are appropriate control inputs to be updated in each period. It can be concluded from Theorems 8.1 that the closed-loop system resulted by the active secure control strategy is equivalent to the case without DoS attacks. Subsequently, the secure controller design procedure has been given in Algorithm 8.2. According to Theorem 8.2, the active secure control approach proposed in this chapter can guarantee that the addressed CPS under two-channel DoS attacks is asymptotically stable without losing the control performance. Finally, both the simulations and experiments were given to show the effectiveness of the proposed active secure control approach.

References

1. Teixeira A, Shames I, Sandberg H, Johansson KH (2015) A secure control framework for resource-limited adversaries. Automatica 51:135–148
2. Ding D, Wang Z, Wei G, Alsaadi FE (2016) Event-based secure control for discrete-time stochastic systems. IET Control Theory Appl 10(15):1808–1815
3. Dolk VS, Tesi P, De Persis C, Heemels WPMH (2017) Event-triggered control systems under denial-of-service attacks. IEEE Trans Control Netw Syst 4(1):93–105
4. Feng S, Tesi P (2017) Resilient control under denial-of-service: robust design. Automatica 79:42–51
5. Sun YC, Yang GH (2018) Periodic event-triggered resilient control for cyber-physical systems under denial-of-service attacks. J Franklin Inst 355(13):5613–5631
6. De Persis C, Tesi P (2015) Input-to-state stabilizing control under denial-of-service. IEEE Trans Autom Control 60(11):2930–2944
7. Feng S, Tesi P (2017) Networked control systems under denial-of-service: co-located vs. remote architectures. Syst Control Lett 108:40–47
8. Lai S, Chen B, Li T, Yu L (2019) Packet-based state feedback control under DoS attacks in cyber-physical systems. IEEE Trans Circuits Syst II Express Briefs 66(8):1421–1425
9. Sun YC, Yang GH (2018) Event-triggered resilient control for cyber-physical systems under asynchronous DoS attacks. Inf Sci 465:340–352
10. Zhang H, Cheng P, Shi L, Chen J (2015) Optimal denial-of-service attack scheduling with energy constraint. IEEE Trans Autom Control 60(11):3023–3028
11. Hespanha JP, Naghshtabrizi P, Xu Y (2007) A survey of recent results in networked control systems. Proc IEEE 95(1):138–162
12. Kouvaritakis B, Rossiter JA, Schuurmans J (2000) Efficient robust predictive control. IEEE Trans Autom Control 45(8):1545–1549
13. Boyd S, El Ghaoui L, Feron E, Balakrishnan V (1994) Linear matrix inequalities in system and control theory. Society for Industrial & Applied Mathematics
14. Kouvaritakis B, Li S, Cannon M (2010) A line search improvement of efficient MPC. Automatica 46(11):1920–1924

Chapter 9
Security Control Against Time-Delay Attacks Using Switched System Approach

9.1 Introduction

In this chapter, the focus is shifted to study the security control problem under time-delay attacks in CPSs. Unlike the DoS attack discussed in Chap. 8, time-delay attacks involve temporarily jamming communication channels to delay the transmission of signal streams without altering the information within the packet [1–5]. Time-delay attacks are less easily perceived than DoS attacks, and they can degrade system performance or even lead to instability, particularly in systems with high real-time requirements [1, 6].

The proposed approach in this chapter employs a switched system framework with a mode-dependent controller to defend against time-delay attacks in CPSs. The key idea is to divide the attack-induced delays into multiple equilong subintervals, each of which is further separated into nominal and uncertain parts. The system is modeled as a discrete-time switched system with norm-bounded uncertainties, and the mode-dependent controller is designed to adapt to the varying attack-induced delays. This involves switching between different subsystems based on the attack-induced delays, where different controller gains are applied to compensate for different delays. As the switching signals are unavailable due to the unpredictable delays, a "packet-based control" framework was introduced in [7, 8], allowing all control signals corresponding to each subsystem to be sent from the controller to the actuator in a single data packet. The appropriate control signal is then extracted and applied at the actuator.

In summary, this chapter proposes a switched system model with norm-bounded uncertainties to characterize time-delay attacks in CPSs. Leveraging a packet-based control strategy, a mode-dependent controller is designed to defend against time-delay attacks, applying different controller gains for varying attack-induced delays. This approach aims to more effectively compensate for attack-induced delays compared with fixed-gain controllers considered in [9–12], resulting in improved control performance. Experimental results from a networked inverted pendulum control system are provided to validate the effectiveness of the proposed method.

B. Chen et al., *Secure Fusion Estimation and Control for Cyber-Physical Systems Under Malicious Attacks*, https://doi.org/10.1007/978-981-97-9396-9_9

9.2 Modeling of the CPSs Under Time-Delay Attacks

Consider the CPS process shown in Fig. 9.1, where the continuous-time plant is described by the following linear system model

$$\dot{x}(t) = A_p x(t) + B_p u(t), \tag{9.1}$$

where $x \in \Re^n$ and $u \in \Re^m$ are the system state and the control input, respectively, while $A_p \in \Re^{n\times n}$ and $B_p \in \Re^{n\times m}$ are system matrices. The sensor is time driven with the sampling period T, whereas the controller and the actuator are event driven. Besides, the control input u in system (9.1) is subjected to the following constraints:

$$-\bar{u}_i \le u_i \le \bar{u}_i, \quad i = 1, 2, \cdots, m, \tag{9.2}$$

where u_i is the ith element of control input u, and $\bar{u}_i$ is a known constant.

In Fig. 9.1, the plant is controlled through a communication network, and the communication channels of the CPSs are subject to time-delay attacks. These attacks involve temporarily jamming the communication channels to delay the transmission of signal streams without altering the information enclosed in the packet. Importantly, the attacker aims to maintain stealthiness by ensuring that the attack-induced delay is less than one sampling period T, as delays larger than one sampling period are more easily detected by the CPSs. Therefore, denoting the attack-induced delay at time k as τ_k, we consider that τ_k is smaller than T and bounded by $0 \le \tau_m \le \tau_k \le \tau_M \le T$.

In this case, there are two control inputs $u(k)$ and $u(k-1)$, which act on the actuator in one sampling period. Then, system (9.1) is discretized to

$$x(k+1) = Ax(k) + B_{c0}(\tau_k)u(k) + B_{c1}(\tau_k)u(k-1), \tag{9.3}$$

where $A = e^{A_p T}$, $B_{c0}(\tau_k) = \int_0^{T-\tau_k} e^{A_p s} B_p ds$ and $B_{c1}(\tau_k) = \int_{T-\tau_k}^{T} e^{A_p s} B_p ds$.

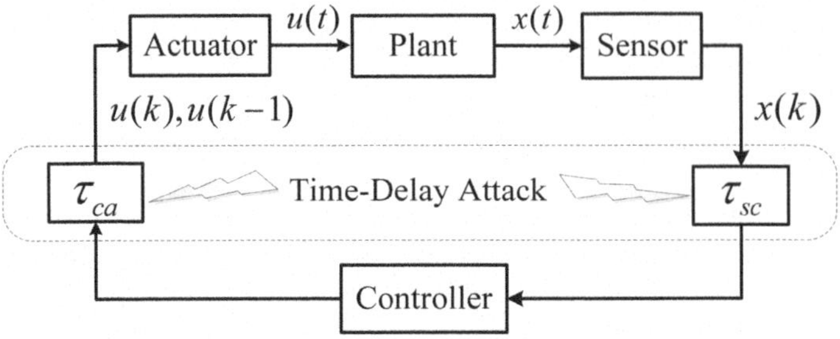

Fig. 9.1 The schematic diagram of the CPSs under time-delay attacks

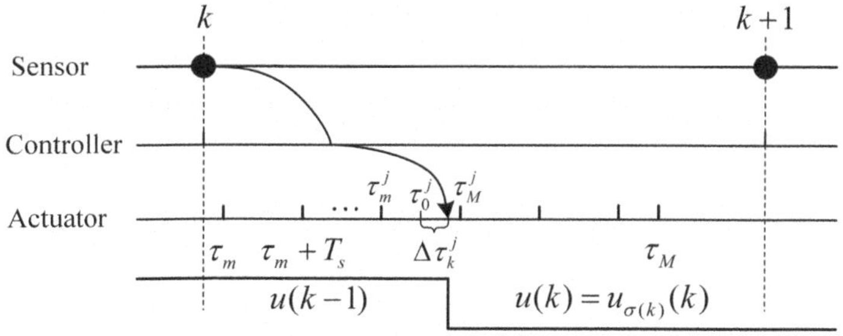

Fig. 9.2 The timing diagram of the CPSs with mode-dependent controller

As shown in Fig. 9.2, we divide interval $[\tau_m, \tau_M]$ into $N+1$ adjacent subintervals and each subinterval $[\tau_m^j, \tau_M^j)$ satisfies

$$\tau_m = \tau_m^1 < \tau_M^1 = \tau_m^2 < \tau_M^2 = \tau_m^3 < \cdots < \tau_M^{N+1} = \tau_M, \tag{9.4}$$

where $N = \lfloor L/T_s \rfloor$, $L = \tau_M - \tau_m$ is the length of $[\tau_m, \tau_M]$ and L_j is the length of $[\tau_m^j, \tau_M^j)$. Let $L_1 = L_2 = \cdots = L_N = T_s$ and $L_{N+1} = L - NT_s$. Then, the unknown attack-induced delay τ_k can be written as

$$\tau_k = \tau_0^j + \Delta\tau_k^j, \tag{9.5}$$

where $\tau_0^j \in [\tau_m^j, \tau_M^j)$ is called the nominal part of τ_k, $\Delta\tau_k^j = \tau_k - \tau_0^j$ is called the uncertain part of τ_k and $\Delta\tau_k^j \in [\tau_m^j - \tau_0^j, \tau_M^j - \tau_0^j)$. Let $Z_0 \triangleq \{1, 2, \cdots, N+1\}$, then $j \in Z_0$.

For convenience, a mapping $\mathfrak{R} \to \mathfrak{R} : \tau_k \to \sigma(k)$ is introduced as follows:

$$\begin{aligned} &\tau_k \in [\tau_m^1, \tau_M^1) \to 1, \\ &\tau_k \in [\tau_m^2, \tau_M^2) \to 2, \\ &\quad\vdots \\ &\tau_k \in [\tau_m^{N+1}, \tau_M^{N+1}] \to N+1, \end{aligned}$$

where $\sigma(k) \in Z_0$.

Define $A_{0\sigma(k)} = e^{A_p(T-\tau_0^j)}$ and $B_{0\sigma(k)} = \int_0^{T-\tau_0^j} e^{A_p s} B_p ds$. Then, it follows from (9.5) that

$$
\begin{aligned}
B_{c0}(\tau_k) &= \int_0^{T-\tau_k} e^{A_p s} B_p ds = \int_0^{T-\tau_0^j-\Delta\tau_k^j} e^{A_p s} B_p ds \\
&= \int_0^{T-\tau_0^j} e^{A_p s} B_p ds + \int_{T-\tau_0^j}^{T-\tau_0^j-\Delta\tau_k^j} e^{A_p s} B_p ds \\
&= \int_0^{T-\tau_0^j} e^{A_p s} B_p ds + e^{A_p(T-\tau_0^j)} \int_0^{-\Delta\tau_k^j} e^{A_p s} B_p ds \\
&= B_{0\sigma(k)} + A_{0\sigma(k)}\Theta_0(\Delta\tau_k^j) B_p,
\end{aligned}
\tag{9.6}
$$

where $\Theta_0(\Delta\tau_k^j) = \int_0^{-\Delta\tau_k^j} e^{A_p s} ds$. Let $B = \int_0^T e^{A_p s} B_p ds$ and $B_{1\sigma(k)} = B - B_{0\sigma(k)}$. Since $B_{c1}(\tau_k) + B_{c0}(\tau_k) = B$, one has

$$
\begin{aligned}
B_{c1}(\tau_k) &= B - B_{c0}(\tau_k) = B - B_{0\sigma(k)} - A_{0\sigma(k)}\Theta_0(\Delta\tau_k^j) B_p \\
&= B_{1\sigma(k)} - A_{0\sigma(k)}\Theta_0(\Delta\tau_k^j) B_p.
\end{aligned}
\tag{9.7}
$$

Then, by (9.6) and (9.7), system (9.3) is rewritten as a discrete-time switched system model with uncertainties:

$$
\begin{aligned}
x(k+1) = Ax(k) &+ \left(B_{0\sigma(k)} + \Delta B_{0\sigma(k)}\right) u(k) \\
&+ \left(B_{1\sigma(k)} - \Delta B_{0\sigma(k)}\right) u(k-1),
\end{aligned}
\tag{9.8}
$$

where $\Delta B_{0\sigma(k)} = A_{0\sigma(k)}\Theta_0(\Delta\tau_k^j) B_p$ and $\sigma(k) \in Z_0$ are the switching signals. Denote $\sigma_{\max}(\Delta\tau_k^j)$ as the maximum singular value of the matrix $\Theta_0(\Delta\tau_k^j)$, and define

$$
\sigma(T, N) = \sup_{\Delta\tau_k^j \in [\tau_m^j - \tau_0^j, \tau_M^j - \tau_0^j)} \sigma_{\max}(\Delta\tau_k^j). \tag{9.9}
$$

Then, it can be obtained that $\|\Theta_0(\Delta\tau_k^j)\|^2 \le \sigma^2(T, N)$, where $\sigma(T, N)$ is a finite scalar since the delay is upper-bounded.

To compensate the attack-induced delays, the following mode-dependent controller is considered for the proposed switched system (9.8)

$$
u(k) = K_{\sigma(k)} x(k), \tag{9.10}
$$

where $\sigma(k)$ is a switching signal determined by mapping $\tau_k \to \sigma(k)$. In other words, the controller is switched on the basis of the attack-induced delays, and different controller gains are applied to compensate different delays.

Based on the controller (9.10) and the proposed switched system (9.8), the closed-loop CPS model is obtained as follows:

$$\begin{aligned} x(k+1) = &\left(A + B_{0\sigma(k)} K_{\sigma(k)} + \Delta B_{0\sigma(k)} K_{\sigma(k)}\right) x(k) \\ &+ \left(B_{1\sigma(k)} K_{\sigma(k-1)} - \Delta B_{0\sigma(k)} K_{\sigma(k-1)}\right) x(k-1). \end{aligned} \tag{9.11}$$

Consequently, the closed-loop CPS is characterized as a discrete-time switched system with norm-bounded uncertainties. The system is presumed to be the transition between different subsystems contingent upon the attack-induced delays falling into distinct subintervals.

Due to the unpredictable attack-induced delay, the switching signal is not available for the controller during the transmission. Therefore, the controller sends all the control signals in one data packet to the actuator. In other words, the control sequence $U(k)$ is packed into one data packet and transmitted to the actuator instead of only transmitting $u(k)$.

$$\begin{aligned} U(k) &= \left[u_1(k) \ \cdots \ u_i(k) \ \cdots \ u_{N+1}(k)\right] \\ &= \left[K_1 x(k) \ \cdots \ K_i x(k) \ \cdots \ K_{N+1} x(k)\right]. \end{aligned} \tag{9.12}$$

Then, the switching signal can be obtained when the packet arrives at actuator. The appropriate control signal is packed out based on the known switching signal and applied to the plant.

Remark 9.1 Notice that the time-delay attacks may not be implemented in CPSs at all times, and the attack-induced delays are usually time-varying. Thus, compared with the robust control approach which is designed based on the bound of delays, the proposed switched system approach with mode-dependent controller is more effective to defend against the time-delay attacks, where different controller gains are applied to compensate different attack-induced delays.

Remark 9.2 In [13], a similar mode-dependent controller is designed by using a special time-driven actuator with a period T_0 no longer than the sampling period T (i.e., $T_0 = T/N$ for some positive integer N). It implies that the special time-driven actuator needs to be adapted to different T and N in CPSs. More importantly, when $T_0 = T/N$ is an infinite decimal (e.g., $T_0 = 0.333\cdots$), such a special time-driven actuator cannot be accurately implemented in practical applications. Therefore, the event-driven actuator considered in this chapter is easy to adapt to different system parameters compared with the time-driven actuator in [11–13].

9.3 Stability Analysis and Controller Design for the CPSs Under Time-Delay Attacks

Before proceeding further, some useful definitions are given as follows.

Definition 9.1 The closed-loop CPS (9.11) is exponentially stable, if for each finite initial state $(x(0), x(1)) \in \Re^n$, there exist positive scalars $\alpha > 0$ and $\beta < 1$ such that the following inequality holds:

$$\|x(k)\| \leq \alpha\beta^k \|x_0\|_m,$$

where $\|x_0\|_m = \max\{\|x(0)\|, \|x(1)\|\}$.

Definition 9.2 For each $k_2 > k_1 \geq 0$, let $T_\sigma[k_1, k_2)$ denote the number of switching points of $\sigma(k)$ during $[k_1, k_2)$. If

$$T_\sigma[k_1, k_2) \leq T_0 + (k_2 - k_1)/t_a$$

holds for $T_0 \geq 0$ and $t_a > 0$, then t_a is called the average dwell time and T_0 is the chatter bound. For simplicity, but without loss of generality, we choose $T_0 = 0$.

9.3.1 Stability Analysis of the CPSs

The stability analysis for the exponential stability of the closed-loop CPS (9.11) under time-delay attacks will be presented in Theorem 9.1.

Theorem 9.1 *Consider the closed-loop CPS* (9.11), *if there exist matrices* $\alpha > 0$, Z, K_j, $P_j > 0$ *and* $Q_j > 0$, $j = 1, 2, \cdots, N+1$ *such that*

$$\Omega_j = \begin{bmatrix} A_{cj}^{\mathrm{T}} \\ B_{cj}^{\mathrm{T}} \end{bmatrix} P_j \begin{bmatrix} A_{cj} & B_{cj} \end{bmatrix} + \begin{bmatrix} -\lambda^2 P_j + K_j^{\mathrm{T}} Q_j K_j & \mathbf{0} \\ \mathbf{0} & -\lambda^2 Q_j \end{bmatrix} < 0, \tag{9.13}$$

$$P_a \leq \mu P_b, Q_a \leq \mu Q_b, \forall a, b \in Z_0, \tag{9.14}$$

$$x^{\mathrm{T}}(0) P_j x(0) \leq \alpha, \tag{9.15}$$

$$\begin{bmatrix} Z & K_j \\ K_j^{\mathrm{T}} & \alpha^{-1} P_j \end{bmatrix} \geq 0, \tag{9.16}$$

$$Z_{ii} \leq \bar{u}_i^2, i = 1, 2, \cdots, m, \tag{9.17}$$

$$\lambda < \mu^{-1/2} \tag{9.18}$$

hold, then the closed-loop CPS (9.11) *under time-delay attacks is exponentially stable with the decay rate* $\rho(\lambda, \mu) \leq \lambda\mu^{1/2}$, *and the* $u(k) = K_j x(k)$ *satisfies the constraint* (9.2)*, where* $\lambda < 1$ *and* $\mu \geq 1$ *are given positive scalars,* Z_{ii} *is the* i*th element on the diagonal of matrix* Z.

Proof Let $k_1, k_2, \cdots, k_s, s \geq 1$ denote the switching time of $\sigma(k)$ during $[0, k)$, and $0 < k_1 < \cdots < k_s < k$. The subsystem of closed-loop CPS (9.11) is

$$x(k+1) = A_{cj}x(k) + B_{cj}K_{j-1}x(k-1), \ j \in Z_0, \tag{9.19}$$

where $A_{cj} = A + B_{0j}K_j + \Delta B_{0j}K_j$, $B_{cj} = B_{1j} - \Delta B_{0j}$.

Choose the Lyapunov candidate function for (9.19) as follows:

$$V_j(k) = x^{\mathrm{T}}(k)P_jx(k) + x^{\mathrm{T}}(k-1)K_{j-1}^{\mathrm{T}}Q_jK_{j-1}x(k-1). \tag{9.20}$$

Let $\xi(k) = \left[x(k), K_{j-1}x(k-1)\right]^{\mathrm{T}}$, it is obtained from (9.13) that

$$V_j(k+1) - \lambda^2 V_j(k) = \xi^{\mathrm{T}}(k)\Omega_j\xi(k) < 0. \tag{9.21}$$

Choose the Lyapunov candidate function for (9.11) as follows:

$$V_{\sigma(k)}(k) = x^{\mathrm{T}}(k)P_{\sigma(k)}x(k) + x^{\mathrm{T}}(k-1)K_{\sigma(k-1)}^{\mathrm{T}}Q_{\sigma(k)}K_{\sigma(k-1)}x(k-1), \tag{9.22}$$

where P_j and Q_j, $j \in Z_0$, are the solutions of (13)-(18). From (9.21) and (9.22), one has

$$\begin{cases} V_{\sigma(k)}(k) = V_{\sigma(k_s)}(k) \leq \lambda^{2(k-k_s)}V_{\sigma(k_s)}(k_s), \\ V_{\sigma(k_j-1)}(k_j) = V_{\sigma(k_{j-1})}(k_j) \leq \lambda^{2(k_j-k_{j-1})}V_{\sigma(k_{j-1})}(k_{j-1}). \end{cases} \tag{9.23}$$

Because the state of CPS (9.11) does not jump at switching points, it follows from (9.14) that

$$\begin{aligned} V_{\sigma(k_j)}(k_j) &= x^{\mathrm{T}}(k_j)P_{\sigma(k_j)}x(k_j) + x^{\mathrm{T}}(k_j-1)K_{\sigma(k_j-1)}^{\mathrm{T}}Q_{\sigma(k_j)}K_{\sigma(k_j-1)}x(k_j-1) \\ &\leq \mu x^{\mathrm{T}}(k_j)P_{\sigma(k_j-1)}x(k_j) + \mu x^{\mathrm{T}}(k_j-1)K_{\sigma(k_j-1)}^{\mathrm{T}}Q_{\sigma(k_j-1)}K_{\sigma(k_j-1)}x(k_j-1) \\ &= \mu V_{\sigma(k_j-1)}(k_j), \ j = 1, 2, \cdots, s. \end{aligned} \tag{9.24}$$

Combining (9.23) and (9.24), one has

$$
\begin{aligned}
V_{\sigma(k)}(k) &\leq \lambda^{2(k-k_s)} V_{\sigma(k_s)}(k_s) \leq \lambda^{2(k-k_s)} \mu V_{\sigma(k_s-1)}(k_s) \\
&\leq \mu \lambda^{2(k-k_s)} \lambda^{2(k_s-k_{s-1})} V_{\sigma(k_{s-1})}(k_{s-1}) \\
&\vdots \\
&\leq \mu^{T_\sigma[0,k)} \lambda^{2(k-k_s)} \lambda^{2(k_s-k_{s-1})} \cdots \lambda^{2k_1} V_{\sigma(0)}(0) \\
&= \mu^{T_\sigma[0,k)} \lambda^{2k} V_{\sigma(0)}(0) \leq \mu^{k/t_a} \lambda^{2k} V_{\sigma(0)}(0) \\
&= \rho(\lambda, t_a)^{2k} V_{\sigma(0)}(0).
\end{aligned} \tag{9.25}
$$

Then, it can be obtained from (9.22) and (9.25) that

$$
\gamma_1 \|x(k)\|^2 \leq V_{\sigma(k)}(k) \leq \rho^{2k} V_{\sigma(0)}(0) \leq \rho^{2k} \gamma_2 \|x_0\|^2,
$$

where $\gamma_2 = \max\limits_{j \in Z_0} \left(\lambda_{\max}\left(P_j\right) + \lambda_{\max}\left(K_{j-1}^{\mathrm{T}} Q_j K_{j-1}\right)\right)$ and $\gamma_1 = \min\limits_{j \in Z_0} \lambda_{\min}\left(P_j\right)$, $\lambda_{\min}(*)$ and $\lambda_{\max}(*)$ are respectively the minimum and maximum eigenvalues of matrices. Then, we further obtain that

$$
\|x(k)\| \leq \sqrt{\gamma_2/\gamma_1} \rho^k \|x_0\|.
$$

Since $t_a \geq 1$, inequality (9.18) and $\lambda < 1$ ensure that $\rho < 1$. Hence, the CPS (9.11) is exponentially stable with a decay rate $\rho(\lambda, \mu) \leq \lambda \mu^{1/2}$.

Applying the Schur complement lemma [15], it can be seen that (9.16) is equivalent to $\alpha K_j P_j^{-1} K_j^{\mathrm{T}} \leq Z$. Let $\left(K_j\right)_i$ denote the i-th row of the matrix K_j, it follows from (9.15) that

$$
\begin{aligned}
|u_i(k)|^2 &= |(K_j)_i x(k)|^2 = |(K_j)_i P_j^{-1/2} P_j^{1/2} x(k)|^2 \\
&\leq \|(K_j)_i P_j^{-1/2}\|^2 \|P_j^{1/2} x(k)\|^2 = (K_j)_i P_j^{-1} (K_j)_i^{\mathrm{T}} x^{\mathrm{T}}(k) P_j x(k) \\
&\leq (K_j)_i P_j^{-1} (K_j)_i^{\mathrm{T}} \alpha \leq Z_{ii}.
\end{aligned} \tag{9.26}
$$

Then, according to (9.17) and (9.26), it can be concluded that $u(k) = K_j x(k)$ satisfies the constraint (9.2). This completes the proof. □

9.3.2 Controller Design

The design procedures of mode-dependent controller are presented in Theorem 9.2.

Theorem 9.2 *Consider the closed-loop CPS* (9.11)*, if there exist matrices* $\omega > 0$*,* Z*,* X_j*,* $Y_j > 0$ *and* $V_j > 0$*,* $j = 1, 2, \cdots, N+1$ *such that* (9.17), (9.18) *and LMIs*

$$\begin{bmatrix} -X_j + \omega A_{0j} A_{0j}^{\mathrm{T}} & AX_j + B_{0j} V_j & B_{1j} Y_j & \mathbf{0} & \mathbf{0} \\ * & -\lambda^2 X_j & \mathbf{0} & V_j^{\mathrm{T}} B_p^{\mathrm{T}} & V_j^{\mathrm{T}} \\ * & * & -\lambda^2 Y_j & -Y_j^{\mathrm{T}} B_p^{\mathrm{T}} & \mathbf{0} \\ * & * & * & -\omega \delta^{-2} \mathbf{I} & \mathbf{0} \\ * & * & * & * & -Y_j \end{bmatrix} < 0, \tag{9.27}$$

$$X_a \leq \mu X_b, Y_a \leq \mu Y_b, \forall a, b \in Z_0, \tag{9.28}$$

$$\begin{bmatrix} \mathbf{I} & U^{\mathrm{T}} \\ U & X_j \end{bmatrix} \geq 0, \tag{9.29}$$

$$\begin{bmatrix} Z & V_j \\ V_j^{\mathrm{T}} & X_j \end{bmatrix} \geq 0 \tag{9.30}$$

hold, then by using the mode-dependent controller $u(k) = V_j X_j^{-1} x(k)$, *the closed-loop CPS* (9.11) *under time-delay attacks is exponentially stable with the decay rate* $\rho(\lambda, \mu) \leq \lambda \mu^{1/2}$, *and* $u(k) = V_j X_j^{-1} x(k)$ *satisfies the constraint* (9.2), *where* $\lambda < 1$, $\mu \geq 1$ *and* $\delta > \sigma(T, N)$ *are positive scalars.*

Proof Utilizing the Schur complement lemma [15], the inequality (9.13) can be converted to

$$\begin{bmatrix} -P_j^{-1} & A_{cj} & B_{cj} \\ * & -\lambda^2 P_j + K_j^{\mathrm{T}} Q_j K_j & \mathbf{0} \\ * & * & -\lambda^2 Q_j \end{bmatrix} < 0, \tag{9.31}$$

which can be written as

$$\Psi_0 + \bar{D} \Theta_0(\Delta \tau_k^j) \bar{E} + \bar{E}^{\mathrm{T}} \Theta_0^{\mathrm{T}}(\Delta \tau_k^j) \bar{D}^{\mathrm{T}} < 0, \tag{9.32}$$

where

$$\Psi_0 = \begin{bmatrix} -P_j^{-1} & A + B_{0j} K_j & B_{1j} \\ * & -\lambda^2 P_j + K_j^{\mathrm{T}} Q_j K_j & \mathbf{0} \\ * & * & -\lambda^2 Q_j \end{bmatrix}, \bar{D} = \begin{bmatrix} A_{0j} \\ \mathbf{0} \\ \mathbf{0} \end{bmatrix}, \bar{E} = \begin{bmatrix} \mathbf{0} & B_p K_j & -B_p \end{bmatrix}.$$

It can be inferred from $\delta > \sigma(T, N)$ that $\Theta_0^{\mathrm{T}}(\Delta \tau_k^j) \Theta_0(\Delta \tau_k^j) < \sigma^2(T, N) < \delta^2$. Therefore, it can be obtained by Lemma 2.2 in [14] that (9.32) is true if and only if there exists a scalar $\varepsilon > 0$ such that

$$\Psi_0 + \varepsilon \bar{D} \bar{D}^{\mathrm{T}} + \varepsilon^{-1} \delta^2 \bar{E}^{\mathrm{T}} \bar{E} < 0 \tag{9.33}$$

holds. Then, applying the Schur complement lemma [15], the inequality (9.33) can be converted to

$$\begin{bmatrix} -P_j^{-1} + \varepsilon A_{0j} A_{0j}^{\mathrm{T}} & A + B_{0j} K_j & B_{1j} & \mathbf{0} \\ * & -\lambda^2 P_j + K_j^{\mathrm{T}} Q_j K_j & \mathbf{0} & K_j^{\mathrm{T}} B_p^{\mathrm{T}} \\ * & * & -\lambda^2 Q_j & -B_p^{\mathrm{T}} \\ * & * & * & -\varepsilon \delta^{-2} \mathbf{I} \end{bmatrix} < 0. \tag{9.34}$$

Denote $\omega = \alpha\varepsilon$, $X_j = \alpha P_j^{-1}$, $Y_j = \alpha Q_j^{-1}$ and $V_j = K_j X_j$. Then, pre- and post-multiplying the inequality (9.34) by $diag\left\{\alpha^{1/2}\mathbf{I}, \alpha^{1/2} P_j^{-1}, \alpha^{1/2} Q_j^{-1}, \alpha^{1/2}\mathbf{I}\right\}$, one has

$$\begin{bmatrix} -X_j + \omega A_{0j} A_{0j}^{\mathrm{T}} & A X_j + B_{0j} V_j & B_{1j} Y_j & \mathbf{0} \\ * & -\lambda^2 X_j + V_j^{\mathrm{T}} \alpha^{-1} Q_j V_j & \mathbf{0} & V_j^{\mathrm{T}} B_p^{\mathrm{T}} \\ * & * & -\lambda^2 Y_j & -Y_j^{\mathrm{T}} B_p^{\mathrm{T}} \\ * & * & * & -\omega \delta^{-2} \mathbf{I} \end{bmatrix} < 0. \tag{9.35}$$

According to the Schur complement lemma [15], (9.35) is equivalent to (9.27). Pre- and post- multiplying $P_a \leq \mu P_b$ by $\alpha^{1/2} P_j^{-1}$ and pre- and post- multiplying $Q_a \leq \mu Q_b$ by $\alpha^{1/2} Q_j^{-1}$ yield that the LMI (9.28) holds.

The initial state is assumed to be an unknown variable satisfying the following set

$$\Phi = \left\{ x(-j) | x(-j) \in \Re^n, x(-j) = U v_j, v_j^{\mathrm{T}} v_j \leq 1, j = 0, 1 \right\}, \tag{9.36}$$

where U is a constant matrix.

By using the Schur complement lemma [15], (9.29) is equivalent to

$$U^{\mathrm{T}} P_j U \leq \alpha \mathbf{I}.$$

Then, it can be obtained from (9.36) that the inequality (9.15) is satisfied. Furthermore. pre- and post- multiplying the inequality (9.16) by $diag\left\{\mathbf{I}, \alpha P_j^{-1}\right\}$ yields the LMI (9.30). This completes the proof. □

Based on Theorem 9.2, the implementation procedures of the proposed switched system approach against time-delay attacks can be summarized by Algorithm 9.1.

Remark 9.3 The smaller λ and μ lead to smaller ρ and t_a^*, and then the inequality (9.18) is easier to satisfy. In fact, the minimal λ and μ can be obtained by using the similar procedures of Algorithm 9.1 in [12], and it is thus omitted.

Algorithm 9.1 .

1: For known τ_m, τ_M and T, select appropriate T_s. Then, determine the closed-loop CPS model (9.11), and calculate $\sigma(T, N)$ by (9.9);
2: Select appropriate parameters $\lambda < 1$, $\mu \geq 1$ and $\delta > \sigma(T, N)$, then determine the controller gains by solving the LMIs in Theorem 9.2;
3: **for** $k = 1$ **to** T_{total} (total number of system periods) **do**
4: At time k, sensor transmits the sampling data to the controller and records the local time as $t_s(k)$;
5: The controller calculates the control sequence $U(k) = \left[u_1(k) \ \cdots \ u_i(k) \ \cdots \ u_{N+1}(k)\right]$ based on the received sampling data. Then, pack $U(k)$ into one data packet and send it to actuator simultaneously;
6: The actuator reads the local time as $t_a(k)$ when receiving the control sequence $U(k)$ and then calculates $\tau_k = t_a(k) - t_s(k)$. Select the appropriate control signal in $U(k)$ by the mapping $\tau_k \to \sigma(k)$ and apply it to the plant;
7: **end for**

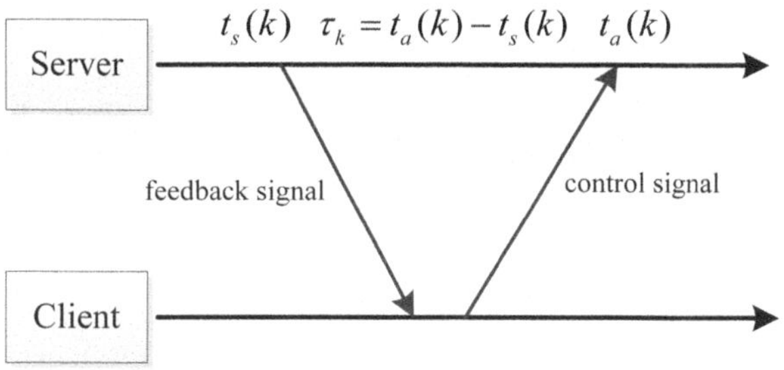

Fig. 9.3 The process of measuring the delay

9.4 Experiment

The networked inverted pendulum control system given in chapter 5 is considered to verify the effectiveness and superiority of the proposed method, as shown in Fig. 9.3. The experiment platform consists of a Client, a Server and a linear inverted pendulum. The system structure diagram can be described as Fig. 9.1. The data packets are transmitted between the Client and the Server over communication network by using the User Datagram Protocol (UDP). The process of measuring the delay is shown in Fig. 9.3, where $t_s(k)$ is the time when the sensor transmits feedback signals to the controller and $t_a(k)$ is the time when the actuator receives control signals. Then, as shown in Algorithm 9.1, the delay can be calculated by $\tau_k = t_a(k) - t_s(k)$ in each sampling period.

The inverted pendulum is an unstable system with nonlinear dynamics. The dynamic equation of this system can be derived based on the second motion law of Newton and the linear state-space equation can be obtained by using classical mechanics theory [16] as follows.

$$\dot{x} = \begin{bmatrix} 0 & 1 & 0 & 0 \\ 0 & 0 & 0 & 0 \\ 0 & 0 & 0 & 1 \\ 0 & 0 & \frac{3g}{4l} & 0 \end{bmatrix} x + \begin{bmatrix} 0 \\ 1 \\ 0 \\ \frac{3}{4l} \end{bmatrix} u, \tag{9.37}$$

where $x = \begin{bmatrix} p & \dot{p} & \psi & \dot{\psi} \end{bmatrix}^{\mathrm{T}}$ and $u = \ddot{p}$. p, $\dot{p}$ and $\ddot{p}$ are the cart displacement, the cart velocity and the cart acceleration, respectively. $\psi = \theta - \pi$, where θ is the angle between pendulum and vertical downward direction, $\dot{\psi}$ and $\ddot{\psi}$ are the angular velocity and the angular acceleration of the pendulum, respectively. $g = 9.8\mathrm{m/s^2}$ is the gravitational acceleration, and $2l = 0.5\mathrm{m}$ is the pendulum length. The control input u of inverted pendulum system (9.37) is $\ddot{p}$, which is subjected to the following constraint:

$$-15m/s^2 \le u \le 15m/s^2. \tag{9.38}$$

Consider the scenario where the bounds of attack-induced delays are $\tau_m = 0.3ms$ and $\tau_M = 3ms$ respectively.

The sensor sampling period is set to $T = 10ms$. Let $T_s = 0.4ms$. The values of the map $\tau_k \to \sigma(k)$ are as follows:

$$\begin{aligned} \tau_k &\in [0.3ms, 0.7ms) \to 1, \\ \tau_k &\in [0.7ms, 1.1ms) \to 2, \\ \tau_k &\in [1.1ms, 1.5ms) \to 3, \\ \tau_k &\in [1.5ms, 1.9ms) \to 4, \\ \tau_k &\in [1.9ms, 2.3ms) \to 5, \\ \tau_k &\in [2.3ms, 2.7ms) \to 6, \\ \tau_k &\in [2.7ms, 3.0ms] \to 7. \end{aligned}$$

By choosing $U = diag\{0.2, 0.2, 0.2, 0.2\}$, $\lambda = 0.997$, $\mu = 1.005$, and utilizing Theorem 9.2, the controller gain matrices are obtained as follows:

$$\begin{aligned} K_1 &= \begin{bmatrix} 1.1402 & 4.0062 & -48.7547 & -8.9919 \end{bmatrix}, \\ K_2 &= \begin{bmatrix} 1.1715 & 4.1157 & -50.1827 & -9.2552 \end{bmatrix}, \\ K_3 &= \begin{bmatrix} 1.2175 & 4.2769 & -52.2659 & -9.6394 \end{bmatrix}, \\ K_4 &= \begin{bmatrix} 1.2853 & 4.5141 & -55.3413 & -10.2066 \end{bmatrix}, \\ K_5 &= \begin{bmatrix} 1.3887 & 4.8745 & -60.1393 & -11.0914 \end{bmatrix}, \\ K_6 &= \begin{bmatrix} 1.5416 & 5.3948 & -68.9713 & -12.7200 \end{bmatrix}, \\ K_7 &= \begin{bmatrix} 1.5665 & 5.4883 & -69.1631 & -12.7555 \end{bmatrix}. \end{aligned}$$

By using Algorithm 9.1, the state trajectories and control inputs in the experiments are depicted in Fig. 9.4. The attack-induced delays and the switching signals are depicted in Figs. 9.5 and 9.6, from which it can be seen that the switching sig-

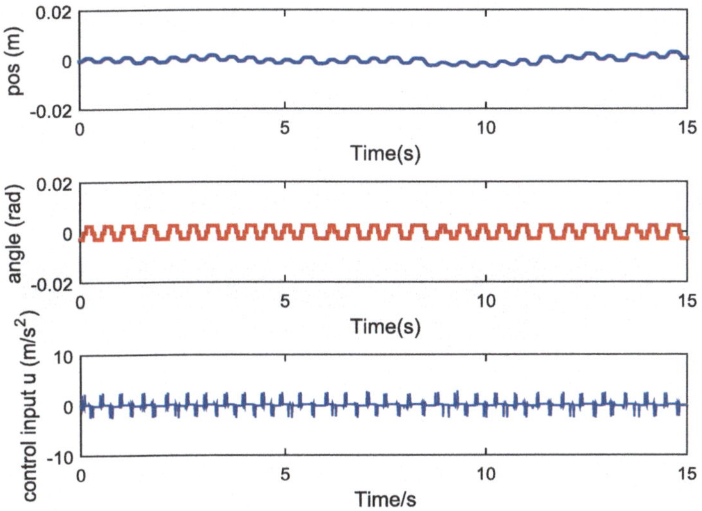

Fig. 9.4 The system outputs and control inputs

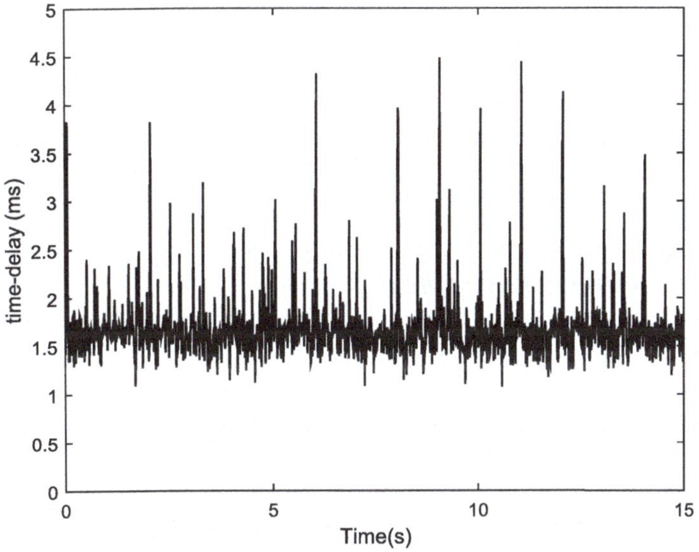

Fig. 9.5 The attack-induced delays

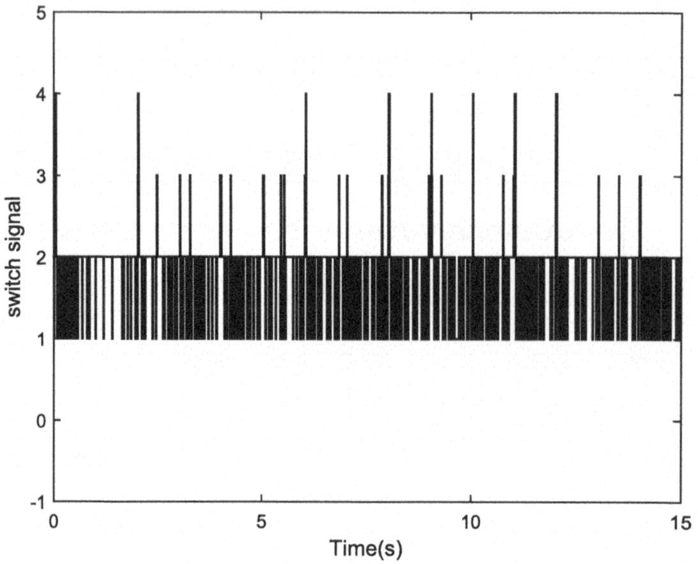

Fig. 9.6 The switch signals

nals depend on the attack-induced delays. Figure 9.4 shows that the system is well stabilized and the control input satisfies the constraint (9.38), which means that our method is valid under the time-delay attacks in practice.

9.4.1 Comparative Experiment

To further verify the superiority of the proposed method, a comparison with a robust control approach [10] is given. Consider the same network environments and set the same parameters as in this chapter. By applying the method in [10], the controller gain can be obtained as follows:

$$K = \begin{bmatrix}1.7072 \; 5.7915 \; -70.0282 \; -12.9151\end{bmatrix}.$$

The comparison between the two methods are depicted in Fig. 9.7. Then, a cumulative error is introduced, which is defined as $\delta(k) = \sum_{i=0}^{k} |e(i)|$, where e is the difference between actual value and expected value. The cumulative errors of state trajectories of the two methods are depicted in Fig. 9.8. From Figs. 9.7 and 9.8, we can conclude that our method is more efficient for this experiment than the method in [10].

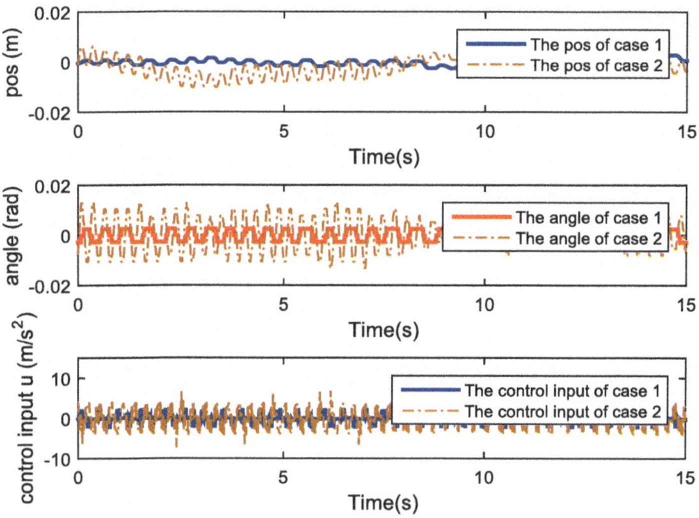

Fig. 9.7 The comparison among the experiments

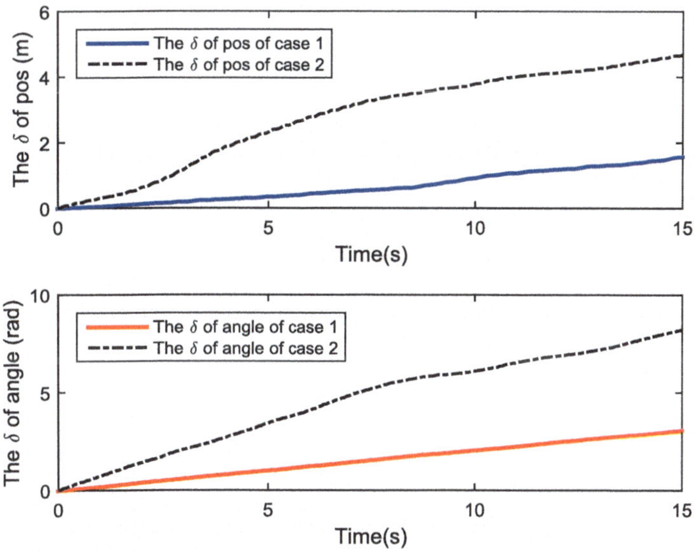

Fig. 9.8 The cumulative error of the position and angle

9.5 Conclusion

A new switched system model was proposed to characterize the time-delay attacks in CPSs. Then, a mode-dependent controller has been designed to ensure the closed-loop CPSs to be exponentially stable, which is switched according to the attack-induced delays and is successfully implemented combined with the packet-based control strategy. Notice that the attack-induced delays can be compensated more effectively by using the mode-dependent controller because different controller gains are applied for different network conditions. Finally, the experiments show that the developed method in this chapter can achieve a better control performance in the real system.

References

1. Koç ÇK (ed) (2018) Cyber-physical systems security, 1st edn. Springer International Publishing
2. Sargolzaei A, Yen KK, Abdelghani MN (2016) Preventing time-delay switch attack on load frequency control in distributed power systems. IEEE Trans Smart Grid 7(2):1176–1185
3. Sargolzaei A, Yen KK, Abdelghani MN, Sargolzaei S (2017) Resilient design of networked control systems under time delay switch attacks, application in smart grid. IEEE Access 5:15901–15912
4. Rahimi K, Parchure A, Centeno V, Broadwater R (2015) Effect of communication time-delay attacks on the performance of automatic generation control. In: 2015 North American power symposium (NAPS)
5. Lou X, Tran C, Tan R, Yau DKY, Kalbarczyk ZT (2019) Assessing and mitigating impact of time delay attack: a case study for power grid frequency control. In: Proceedings of the 10th ACM/IEEE International conference on cyber-physical systems
6. Xu X, Gu GY, Xiong Z, Sheng X, Zhu X (2017) Development of a decentralized multi-axis synchronous control approach for real-time networks. ISA Trans 68:116–126
7. Lai S, Chen B, Li T, Yu L (2019) Packet-based state feedback control under DoS attacks in cyber-physical systems. IEEE Trans Circ Syst II Express Briefs 66(8):1421–1425
8. Zhao YB, Liu GP, Rees D (2009) Design of a packet-based control framework for networked control systems. IEEE Trans Control Syst Technol 17(4):859–865
9. Zhang WA, Yu L (2010) A robust control approach to stabilization of networked control systems with short time-varying delays. Acta Autom Sin 36(1):87–91
10. Zhang H, Zhang Z, Wang Z, Shan Q (2016) New results on stability and stabilization of networked control systems with short time-varying delay. IEEE Trans Cybernet 46(12):2772–2781
11. Zhang WA, Yu L, Song HB (2008) A switched system approach to networked control systems with time-varying delays. In: 2008 27th Chinese control conference
12. Zhang WA, Yu L, Yin S (2011) A switched system approach to H_∞ control of networked control systems with time-varying delays. J Franklin Inst 348(2):165–178
13. Li T, Zhang WA, Yu L (2019) Improved switched system approach to networked control systems with time-varying delays. IEEE Trans Control Syst Technol 27(6):2711–2717
14. Wang Y, Xie L, de Souza CE (1992) Robust control of a class of uncertain nonlinear systems. Syst Control Lett 19(2):139–149

15. Boyd S, El Ghaoui L, Feron E, Balakrishnan V (1994) Linear matrix inequalities in system and control theory. Society for Industrial & Applied Mathematics
16. Prasad LB, Tyagi B, Gupta HO (2012) Modelling and simulation for optimal control of nonlinear inverted pendulum dynamical system using PID controller and LQR. In: 2012 Sixth Asia modelling symposium

Chapter 10
Adaptive Output Regulation for CPSs Under Time-Delay Attacks

10.1 Introduction

In Chap. 8, the security control algorithm for addressing time-delay attacks is developed using the switching system method, which relies on the knowledge of the system model. Nevertheless, if the system model is unknown, the aforementioned approach becomes inapplicable. In the existing literature, various approaches have been put forward to tackle delays in CPSs. These approaches include the random method [1, 2], the robust method [3, 4], and the switched system method [5, 6]. While these methods effectively address delays, it is important to note that many of these approaches rely on having precise knowledge of the system model. Recently, model-free methods have been increasingly designed in the field of control [7–14]. Reinforcement learning (RL) [15], a technique that learns optimal behaviors through observing and interacting with an environment using sub-optimal control policies [16], offers a significant advantage in tackling model-free problems. Adaptive dynamic programming (ADP), an important subfield of RL [17], serves as an effective approach for solving optimal control problems. For instance, in [7–9], the RL and ADP methods were employed to tackle optimal control problems in unknown linear systems. These works specifically addressed single-channel delays, constant delays in multiple channels, and time-varying delays in multiple channels, respectively. Although measurements were utilized to learn the optimal controllers in these studies, the aspect of disturbance rejection was not taken into consideration. The control problem of securing unknown CPSs against actuator attacks and unmatched perturbations was addressed in [17]. Although a model-free ADP algorithm was designed, the neural networks were introduced to approximate the value function and the identification error was added in the process of designing. The robust resilient control problem of a partially unknown nonlinear system under actuator attacks was tackled in [14] using ADP techniques. Furthermore, the ADP method was also applied to solve the attack scheduling problem [18], the fault-tolerant problem [19] and privacy protection problem [20]. However, to the best knowledge of the authors, there is limited research available on securing unknown CPSs against time-delay attacks.

B. Chen et al., *Secure Fusion Estimation and Control for Cyber-Physical Systems Under Malicious Attacks*, https://doi.org/10.1007/978-981-97-9396-9_10

Based on the analysis above, this chapter focuses on addressing the adaptive output regulation problem for unknown CPSs under time-delay attacks in the S-C channel and the C-A channel. A measurement feedback controller is designed only using control inputs and tracking errors. The model-free RL method is adopted to update the control gains online. While the method of designing the controller bears similarity to the one proposed in [7], there are differences in terms of the location of the delay, and attack-induced delays exist in two channels. Based on the measurement feedback controller, the closed-loop system could achieve both asymptotic tracking and disturbance rejection.

10.2 Problem Statement

In this section, the model of the CPS under time-delay attacks is obtained, and the augmented error system with measurement data is established. Then, the optimal output regulation of the CPS is described.

10.2.1 System Model with Attack-Induced Delays

Consider the following discrete-time system with unknown parameters:

$$\begin{cases} x(k+1) = Ax(k) + Bu(k) + Dr(k), \\ e(k) = Cx(k) - Fr(k), \end{cases} \tag{10.1}$$

where $x(k) \in \mathfrak{R}^n$ is the state vector, $u(k) \in \mathfrak{R}^1$ is the input vector, $r(k) \in \mathfrak{R}^q$ is the exostate vector, $e(k) \in \mathfrak{R}$ is the measurement output, and $A \in \mathfrak{R}^{n\times n}$, $B \in \mathfrak{R}^{n\times 1}$, $C \in \mathfrak{R}^{1\times n}$, $D \in \mathfrak{R}^{n\times q}$, $F \in \mathfrak{R}^{1\times q}$ are all unknown constant matrices. The exostate vector $r(k)$ is generated by the following exosystem:

$$r(k+1) = Er(k), \tag{10.2}$$

where $E \in \mathfrak{R}^{q\times q}$ is an unknown Hurwitz matrix. This exosystem generates both the disturbance $d(k) = Dr(k)$ and the reference signal $w(k) = Fr(k)$ to the system. In fact, the exosystem is a description together with the disturbance and the reference signal, the specific origin of the system (10.1) and the exosystem (10.2) can be referred to the literature [21]. In this chapter, the reference signal is unknown and the tracking error $e(k) \in \mathfrak{R}$ is the solely measurable output variable. Take the Internet of Vehicles as an example, the tracking errors can be measured by installing the sensors on the follower cars, but the reference trajectory of the leader car is unknown. Furthermore, according to the CPS structure shown in Fig. 10.1, when the system output $y(k) = Cx(k)$ and the control input $u(k)$ are sent to the controller and the

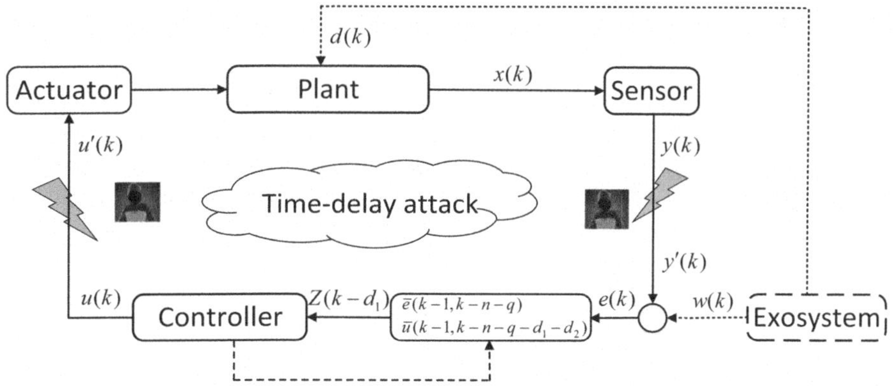

Fig. 10.1 The Frame of CPSs under attacks

actuator through communication networks, respectively, the S-C channel or C-A channel may suffer from the adversary time-delay attacks. The model of the time-delay attack can be described as

$$\begin{cases} u'(k) = u(k - d_{1,k}), \\ y'(k) = y(k - d_{2,k}), \end{cases} \tag{10.3}$$

where $d_{1,k} \geq 0$ and $d_{2,k} \geq 0$ are the attack-induced delays in the C-A channel and S-C channel, respectively. Due to the constraint of attacks' cost, the dwell time of attacks is finite. Hence, the attack-induced delays are assumed to be bounded reasonably, which gives the following assumption.

Assumption 10.1: The delays induced by time-delay attacks are bounded.

From Assumption 10.1, we have $0 \leq d_{1,k} \leq d_1$ and $0 \leq d_{2,k} \leq d_2$, where d_1 and d_2 are the maximal attack-induced delays in the C-A channel and S-C channel, respectively. It should be mentioned that the reference signal can be transmitted through the network or not in practice. Here the security issue is that the attack is launched in the feedback or forward channel of the system, and the system output tracks a given reference signal, i.e., the attack to the reference signal is not considered. For uncertain delays, the robust method is adopted in this chapter, and the bounds of attack-induced delays are utilized to design the controller. More details are explained in Remark 10.1. Based on the bounds of the attack-induced delays, the model of the system (10.1) under attacks can be described as follows:

$$\begin{cases} x(k+1) = Ax(k) + Bu(k - d_1) + Dr(k), \\ e(k) = Cx(k - d_2) - Fr(k). \end{cases} \tag{10.4}$$

Next, the measurement data will be used to reconstruct the system model under time-delay attacks.

10.2.2 Augmented Error System with Measurement Data

As the system state is unmeasurable, designing the controller using direct state feedback is not feasible. Therefore, an alternative approach is adopted, which involves leveraging measurement information to design a measurement feedback controller. This is achieved by establishing the relationship between the system state and the available measurement data. In addition, let us define $\xi(k) = [x^T(k), r^T(k)]^T$, then it follows from (10.2) and (10.4) that the augmented error system is written as

$$\begin{cases} \xi(k+1) = A_1\xi(k) + B_1 u(k-d_1), \\ e(k) = C_1\xi(k-d_2), \end{cases} \tag{10.5}$$

where $A_1 = \begin{bmatrix} A & D \\ \mathbf{0} & E \end{bmatrix}$, $B_1 = \begin{bmatrix} B \\ \mathbf{0} \end{bmatrix}$, and $C_1 = [C, -FE^{d_2}]$.

Assumption 10.2: (A, B) is controllable, and (C_1, A_1) is observable.

Assumption 10.3: The rank of the matrix $\begin{bmatrix} A-\lambda\mathbf{I} & B \\ C & \mathbf{0} \end{bmatrix}$ is $n+1$ for any $\lambda \in \sigma(E)$.

To design a measurement feedback controller, it is necessary to extend the augmented system (10.5) to a system with the measurement data. In the time horizon $[k-n-q-d_1-d_2, k+d_1]$, by using the recursive method, the system (10.5) can be rewritten as

$$\begin{cases} \xi(k+d_1) = \mathcal{A}_1\xi(k-n-q-d_2) + \mathcal{C}_1\bar{u}(k), \\ \bar{e}(k) = \mathcal{O}\xi(k-n-q-d_2) + \mathcal{T}\bar{u}(k), \end{cases} \tag{10.6}$$

where

$$\bar{u}(k) = [u(k-1), \ldots, u(k-n-q-d_1-d_2)]^T,$$
$$\bar{e}(k) = [e(k-1), \ldots, e(k-n-q)]^T,$$

and

$$\mathcal{A}_1 = A_1^{n+q+d_1+d_2},$$
$$\mathcal{C}_1 = [B_1, A_1B_1, \ldots, A_1^{n+q+d_1+d_2-1}B_1],$$
$$\mathcal{O} = [(C_1{A_1}^{n+q-1})^T, \ldots, (C_1A_1)^T, C_1^T]^T,$$
$$\mathcal{T} = \begin{bmatrix} \mathbf{0}_{1\times(d_1+d_2+1)} & C_1B_1 & \ldots & C_1A_1^{n+q-2}B_1 \\ \vdots & \vdots & \ddots & \vdots \\ \mathbf{0}_{1\times(d_1+d_2+1)} & \mathbf{0} & \ldots & C_1B_1 \\ \mathbf{0}_{1\times(d_1+d_2+1)} & \mathbf{0} & \ldots & \mathbf{0} \end{bmatrix}.$$

Since (C_1, A_1) is observable, the observability matrix $\mathcal{O}$ is of full column rank. Define $\mathcal{O}^- = (\mathcal{O}^T\mathcal{O})^{-1}\mathcal{O}^T$, then $\xi(k+d_1)$ is rewritten as

$$\xi(k+d_1) = \mathcal{A}_1\mathcal{O}^-\bar{e}(k) + (\mathcal{C}_1 - \mathcal{A}_1\mathcal{O}^-\mathcal{T})\bar{u}(k) \triangleq M_1 Z(k), \tag{10.7}$$

where

$$M_1 = [\mathcal{A}_1\mathcal{O}^-, \mathcal{C}_1 - \mathcal{A}_1\mathcal{O}^-\mathcal{T}], \tag{10.8}$$

$$Z(k) = [\bar{e}^T(k), \bar{u}^T(k)]^T. \tag{10.9}$$

Then, $x(k+d_1)$ and $r(k+d_1)$ can be expressed uniquely by the measurement variable $Z(k)$, i.e.,

$$\begin{bmatrix} x(k+d_1) \\ r(k+d_1) \end{bmatrix} = M_1 Z(k) \triangleq \begin{bmatrix} M_{1x} \\ M_{1r} \end{bmatrix} Z(k), \tag{10.10}$$

where $M_{1x} \in \Re^{n\times n_z}$, $M_{1r} \in \Re^{q\times n_z}$, $n_z = 2(n+q) + d_1 + d_2$. It follows from the above derivation that $\xi(k+d_1)$ traces back to the state of $\xi(k-n-q-d_2)$. Actually, the historical time $k-n-q-d_2$ can be extended to a general time point $k-N$, as long as N is bigger than $n+q+d_2$. From the similar derivation of $\xi(k+d_1)$, $\xi(k+d_1-d_2)$ has the following form

$$\xi(k+d_1-d_2) = \mathcal{A}_2\xi(k-n-q-d_2) + \mathcal{C}_2\bar{u}(k) \triangleq M_2 Z(k),$$

where $\mathcal{A}_2 = A_1^{n+q+d_1}$, $\mathcal{C}_2 = [\mathbf{0}_{(n+q)\times d_2}, B_1, A_1B_1, \ldots, A_1^{n+q+d_1-1}B_1]$, and $M_2 = [\mathcal{A}_2\mathcal{O}^-, \mathcal{C}_2 - \mathcal{A}_2\mathcal{O}^-\mathcal{T}]$. Following the similar derivation, $x(k+d_1-d_2)$ and $r(k+d_1-d_2)$ can be expressed uniquely by the measurement variable $Z(k)$, i.e.,

$$\begin{bmatrix} x(k+d_1-d_2) \\ r(k+d_1-d_2) \end{bmatrix} = M_2 Z(k) \triangleq \begin{bmatrix} M_{2x} \\ M_{2r} \end{bmatrix} Z(k). \tag{10.11}$$

The dimensions of M_{2x} and M_{2r} are the same with the ones of M_{1x} and M_{1r}, respectively.

10.2.3 Linear Output Regulator

For the linear output regulator problem, the control input can be expressed as the standard form $u(k) = -Kx(k) + (KX + U)r(k)$ [22], where K is obtained by solving an algebraic Riccati equation (ARE), while X and U are obtained by solving regulator equations. Inspired by the standard form of the controller, the following measurement feedback controller will be designed in this chapter:

$$u(k) = [-K(M_{1x} - XM_{1r}) + UM_{1r}]Z(k) \triangleq -\tilde{K}Z(k), \tag{10.12}$$

where K satisfies that $A - BK$ is a Schur matrix, $X \in \mathfrak{R}^{n\times q}$ and $U \in \mathfrak{R}^{m\times q}$ are obtained by solving the following regulator equations:

$$AX + BU + D = XE,$$
$$CX - FE^{d_2} = \mathbf{0}.$$

From Assumption 10.3, the rank of matrix $\begin{bmatrix} A - \lambda\mathbf{I} & B \\ C & \mathbf{0} \end{bmatrix}$ is $n + 1$ for any $\lambda \in \sigma(E)$, and the regulator equations have the unique solutions of X and U. Let

$$\bar{x}(k) = x(k) - Xr(k),$$
$$\bar{u}(k - d_1) = u(k - d_1) - Ur(k).$$

Then the controller (10.12) is equivalent to

$$\bar{u}(k - d_1) = -K\bar{x}(k), \tag{10.13}$$

and the system (10.1) can be rewritten as

$$\begin{cases} \bar{x}(k+1) = (A - BK)\bar{x}(k), \\ e(k) = C\bar{x}(k - d_2). \end{cases} \tag{10.14}$$

Since $A - BK$ is a Schur matrix, one has $\lim_{k\to\infty} \bar{x}(k) = 0$. Then $\lim_{k\to\infty} \bar{u}(k) = 0$, $\lim_{k\to\infty} e(k) = 0$ by (10.13) and (10.14). From the analysis above, the linear output regulation problem can be solved by applying the measurement feedback controller (10.12). At the same time, the open-loop form of the system (10.14) can be expressed as

$$\begin{cases} \bar{x}(k+1) = A\bar{x}(k) + B\bar{u}(k - d_1), \\ e(k) = C\bar{x}(k - d_2). \end{cases} \tag{10.15}$$

In this chapter, the method to find the optimal controller (10.12) for the system under time-delay attacks in both channels will be addressed. The optimality is achieved by minimizing a predefined performance index or value function, which is chosen as

$$J(k) = \sum_{i=k}^{\infty} \left[e^T(i)Qe(i) + \bar{u}^T(i - d_1)R\bar{u}(i - d_1)\right], \tag{10.16}$$

where $Q = Q^T \geq \mathbf{0}$ and $R = R^T > \mathbf{0}$.

So far, the optimal output regulation for CPSs in the presence of attack-induced delays can be converted to solving the following optimization problem:

$$\min J(k), \quad s.t.(10.15). \tag{10.17}$$

Remark 10.1 The measurement feedback controller in this chapter is designed according to the bounds of delays d_1 and d_2, which are related to the dimension of the controller gain. The designed controller is applied to the system where the attack-induced delays are less than the bounds, which means that the control components are redundant. In practice, this control scheme can be achieved through the time stamp at each data segment and a trigger which is introduced artificially in the control/actuator center. Take the S-C channel as an example, when the control center receives the data packet with time information k at time $k + d_{2,k}$ ($d_{2,k} \leq d_2$), the artificial trigger in the control center will not be trigged at once until $k + d_2$, and the data packet with time information k will be adopted at $k + d_2$. If there are still other data packets arrived before the time $k + d_2$, those packets will wait in the buffer. In this case, the proposed method can still deal with the case of time-varying delays.

Remark 10.2 In practice, it is usually hard to obtain an exact system model, making it challenging to design a secure controller. Fortunately, RL, a model-free learning technique, is effective for updating the control policy and performance index based on the control inputs and tracking errors. In this chapter, RL is utilized to continuously update the controller gain, enabling the control strategy to adapt to changing system dynamics and uncertainties in the system model.

10.3 Measurement Feedback Design with Unknown Model Knowledge

In this section, a measurement feedback controller is designed to solve the problem (10.17) with unknown system parameters A, B, C, D, unknown exosystem parameters E, F, unmeasurable system state $x(k)$, exosystem state $r(k)$, and only the control inputs and tracking errors are utilized.

10.3.1 Controller Design

Before presenting the algorithms, the following lemma on the value function is introduced.

Lemma 10.1 *For the infinite-horizon output regulation problem, the reference signal is produced by the exosystem (10.2). Then the value function (10.16) under any fixed stabilizing controller (10.13) can be expressed as*

$$J(k) = \bar{x}^T(k)P\bar{x}(k) \tag{10.18}$$

for some matrix $P = P^T > \mathbf{0}$.

Proof Substituting (10.13) and (10.15) into (10.16) yields

$$
\begin{aligned}
J(k) &= \sum_{i=k}^{\infty} \left(\bar{x}^T(i-d_2)Q_1\bar{x}(i-d_2) + \bar{x}(i)^T K^T RK\bar{x}(i)\right) \\
&= \sum_{i=0}^{\infty} \left(\bar{x}^T(i+k-d_2)Q_1\bar{x}(i+k-d_2) + \bar{x}^T(i+k)K^T RK\bar{x}(i+k)\right),
\end{aligned}
\tag{10.19}
$$

where $Q_1 = C^T QC$. Notice that, based on (10.13), the solution of the system dynamics (10.15) for a specific initial condition $\bar{x}(k)$ is

$$
\bar{x}(i+k-d_2) = (A-BK)^{i-d_2}\bar{x}(k), \tag{10.20}
$$

$$
\bar{x}(i+k) = (A-BK)^{i}\bar{x}(k). \tag{10.21}
$$

Then, it follows from (10.19)–(10.21) that

$$
J(k) = \bar{x}^T(k)P\bar{x}(k),
$$

where

$$
P = \sum_{i=k}^{\infty}\{[(A-BK)^{(i-d_2)}]^T Q_1(A-BK)^{i-d_2} + [(A-BK)^i]^T K^T RK(A-BK)^i\}.
$$

This completes the proof. □

According to (10.18), the value function (10.16) can be rewritten as

$$
\begin{aligned}
\bar{x}^T(k)P\bar{x}(k) = {} & e^T(k)Qe(k) + \bar{u}^T(k-d_1)R\bar{u}(k-d_1) \\
& + \bar{x}^T(k+1)P\bar{x}(k+1).
\end{aligned}
\tag{10.22}
$$

Define $M_{1\bar{x}} = M_{1x} - XM_{1r}$, $M_{2\bar{x}} = M_{2x} - XM_{2r}$. Then

$$
\bar{x}(k) = x(k) - Xr(k) = M_{1\bar{x}}Z(k-d_1), \tag{10.23}
$$

$$
\bar{x}(k-d_2) = M_{2\bar{x}}Z(k-d_1). \tag{10.24}
$$

Substituting (10.15), (10.23) and (10.24) into (10.22) yields

$$
\begin{aligned}
\bar{x}^T(k)P^j\bar{x}(k) &= \bar{x}^T(k-d_2)Q_1\bar{x}(k-d_2) + \bar{u}(k-d_1)^T R\bar{u}(k-d_1) \\
&\quad + (A\bar{x}(k) + B\bar{u}(k-d_1))^T P^j(A\bar{x}(k) + B\bar{u}(k-d_1)) \\
&= Z^T(k-d_1)(M_{2\bar{x}}^T Q_1 M_{2\bar{x}} + M_{1\bar{x}}^T A^T P^j AM_{1\bar{x}})Z(k-d_1) \\
&\quad + Z^T(k-d_1)M_{1\bar{x}}^T A^T P^j B\bar{u}(k-d_1) \\
&\quad + \bar{u}^T(k-d_1)B^T P^j AM_{1\bar{x}}Z(k-d_1) \\
&\quad + \bar{u}^T(k-d_1)(R + B^T P^j B)\bar{u}(k-d_1).
\end{aligned}
$$

Then, by combining the definition of $\bar{u}(k-d_1)$ and (10.10), one has

$$\bar{x}^T(k)P^j\bar{x}(k) = \begin{bmatrix} Z(k-d_1) \\ u(k-d_1) \end{bmatrix}^T \tilde{P}^j \begin{bmatrix} Z(k-d_1) \\ u(k-d_1) \end{bmatrix}, \tag{10.25}$$

where

$$\tilde{P}^j = \begin{bmatrix} \tilde{P}^j_{ZZ} & \tilde{P}^j_{Zu} \\ \tilde{P}^j_{uZ} & \tilde{P}^j_{uu} \end{bmatrix}$$

and

$$\begin{aligned}
\tilde{P}^j_{ZZ} =& M^T_{2\bar{x}} Q_1 M_{2\bar{x}} + M^T_{1\bar{x}} A^T P^j A M_{1\bar{x}} \\
& - M^T_{1\bar{x}} A^T P^j B U M_{1r} - M^T_{1r} U^T B^T P^j A M_{1\bar{x}} \\
& + M^T_{1r} U^T (R + B^T P^j B) U M_{1r} \\
\tilde{P}^j_{Zu} =& M^T_{1\bar{x}} A^T P^j B - M^T_{1r} U^T (R + B^T P^j B) \\
\tilde{P}^j_{uZ} =& B^T P^j A M_{1\bar{x}} - (R + B^T P^j B) U M_{1r} \\
\tilde{P}^j_{uu} =& R + B^T P^j B.
\end{aligned}$$

From the above derivation, the value function (10.22) can be expressed by the measurement data as follows:

$$\begin{aligned}
& \begin{bmatrix} Z(k-d_1) \\ u(k-d_1) \end{bmatrix}^T \tilde{P}^j \begin{bmatrix} Z(k-d_1) \\ u(k-d_1) \end{bmatrix} \\
& = e^T(k) Q e(k) + \bar{u}^T(k-d_1) R \bar{u}(k-d_1) \\
& + \begin{bmatrix} Z(k-d_1+1) \\ -\tilde{K}^j Z(k-d_1+1) \end{bmatrix}^T \tilde{P}^j \begin{bmatrix} Z(k-d_1+1) \\ -\tilde{K}^j Z(k-d_1+1) \end{bmatrix},
\end{aligned} \tag{10.26}$$

Meanwhile, it follows from (10.12) that the measurement feedback controller gain $\tilde{K}^{j+1}$ can be obtained from $\tilde{P}^j$, i.e.,

$$\begin{aligned}
\tilde{K}^{j+1} &= K^{j+1} M_{1\bar{x}} - U M_{1v} \\
&= (R + B^T P^j B)^{-1} \times [B^T P^j A M_{1\bar{x}} - (R + B^T P^j B) U M_{1r}] \\
&= (\tilde{P}^j_{uu})^{-1} \tilde{P}^j_{uZ},
\end{aligned} \tag{10.27}$$

where $K^{j+1} = (R + B^T P^j B)^{-1} B^T P^j A$ is obtained iteratively by solving the optimization problem (10.17) with no time delays, i.e., $d_1 = d_2 = 0$. Then the system (10.15) is reduced to

$$\begin{cases} \bar{x}(k+1) = A\bar{x}(k) + B\bar{u}(k), \\ e(k) = C\bar{x}(k). \end{cases} \tag{10.28}$$

Obviously, it is a standard linear quadratic regulator problem, and the optimal state-feedback controller has the form of

$$\bar{u}^*(k) = -K^*\bar{x}(k),$$

where the controller gain K^* is

$$K^* = (R + B^T P^* B)^{-1} B^T P^* A, \tag{10.29}$$

and the constant matrix P^* satisfies the discrete-time ARE

$$A^T P^* A - P^* + Q_1 - A^T P^* B(R + B^T P^* B)^{-1} B^T P^* A = \mathbf{0}. \tag{10.30}$$

Remark 10.3 Notice that the ARE (10.30) is nonlinear, and much expensive computation is required to obtain P^*. To reduce the computational cost, the policy iteration (PI) algorithm and the value iteration (VI) algorithm, which are two iterative algorithms in RL, are usually employed to approach P^* approximately. When the time goes to infinity, P^j and K^j will reach P^* and K^* with small neighborhoods, respectively. In this case, $u^*(k) = (-K^* M_{1\bar{x}} + U M_{1r})Z(k) = -\tilde{K}^* Z(k)$, and it is equivalent to $\bar{u}^*(k - d_1) = -K^*\bar{x}(k)$. This implies that the measurement feedback controller can stabilize the system (10.15) in an optimal sense.

Next, the measurement data will be used to learn the controller gain $\tilde{K}^*$. Let

$$\begin{aligned}
\psi^j(k) = &\begin{bmatrix} Z(k-d_1) \\ u(k-d_1) \end{bmatrix}^T \otimes \begin{bmatrix} Z(k-d_1) \\ u(k-d_1) \end{bmatrix} \\
&- \begin{bmatrix} Z(k-d_1+1) \\ -\tilde{K}^j Z(k-d_1+1) \end{bmatrix}^T \otimes \begin{bmatrix} Z(k-d_1+1) \\ -\tilde{K}^j Z(k-d_1+1) \end{bmatrix}, \\
\eta(k) = &\, e^T(k) Q e(k) + \bar{u}^T(k-d_1) R \bar{u}(k-d_1).
\end{aligned}$$

Then, (10.26) can be rewritten briefly:

$$\psi^j(k) \cdot vec(\tilde{P}^j) = \eta(k). \tag{10.31}$$

To ensure the solvability of $\tilde{P}^j$, s groups of data will be chosen

$$\begin{bmatrix} \psi^j(k_0 - d_1) \\ \vdots \\ \psi^j(k_0 + s - d_1) \end{bmatrix}^T \cdot vec(\tilde{P}^j) = \begin{bmatrix} \eta(k_0) \\ \vdots \\ \eta(k_0 + s) \end{bmatrix}, \tag{10.32}$$

where $k_0 \geq n + q + d_1$. Define $\Psi^j = [(\psi^j(k_0 - d_1))^T, \ldots, (\psi^j(k_0 + s - d_1))^T]^T$, $\Xi = [\eta^T(k_0), \ldots, \eta^T(k_0 + s)]^T$. Then, (10.32) is described by

Algorithm 10.1 Online PI Algorithm Based on Measurement Data

1: Give an initial stable control policy $\hat{u}(k) = u(k) + \hat{e}(k)$ as the input on the time interval $[k_0, k_0 + s]$, where $\hat{e}(k)$ is the exploration noise. Compute Ξ and $Z(k)$ in this interval;
2: Choose a stabilizing controller gain $\tilde{K}^0$ and the iteration index $j = 0$. Select a constant $\epsilon > 0$.
3: **Policy evaluation**: solve for $\tilde{P}^j$ from (10.33);
4: **Policy improvement**: solve for $\tilde{K}^{j+1}$ from (10.34);
5: Repeat Step 3 and Step 4 until $\|\tilde{P}^j - \tilde{P}^{j-1}\| < \epsilon$.

$$\Psi^j \cdot vec(\tilde{P}^j) = \Xi. \tag{10.33}$$

Thus, when $\tilde{P}^j$ is obtained from (10.33), the controller gain can be updated by

$$\tilde{K}^{j+1} = (\tilde{P}^j_{uu})^{-1}\tilde{P}^j_{uZ}. \tag{10.34}$$

Then, the online PI algorithm via the measurement feedback can be summarized by Algorithm 10.1.

Note that the initial control policy needs to be admissible in the PI algorithm. To avoid the restriction, the VI algorithm is derived later. Firstly, it is derived from (10.22) that

$$\begin{aligned}
&\begin{bmatrix} Z(k-d_1) \\ u(k-d_1) \end{bmatrix}^T \tilde{P}^{j+1} \begin{bmatrix} Z(k-d_1) \\ u(k-d_1) \end{bmatrix} \\
&= e^T(k)Qe(k) + \bar{u}^T(k-d_1)R\bar{u}(k-d_1) \\
&+ \begin{bmatrix} Z(k+1-d_1) \\ -\tilde{K}^j Z(k+1-d_1) \end{bmatrix}^T \tilde{P}^j \begin{bmatrix} Z(k+1-d_1) \\ -\tilde{K}^j Z(k+1-d_1) \end{bmatrix}.
\end{aligned} \tag{10.35}$$

For $j = 0, 1, 2, \ldots$, let

$$\begin{aligned}
\theta(k) &= \begin{bmatrix} Z(k-d_1) \\ u(k-d_1) \end{bmatrix}^T \otimes \begin{bmatrix} Z(k-d_1) \\ u(k-d_1) \end{bmatrix}^T, \\
\phi^j(k) &= e^T(k)Qe(k) + \bar{u}^T(k-d_1)R\bar{u}(k-d_1) \\
&\quad + \begin{bmatrix} Z(k+1-d_1) \\ -\tilde{K}^j Z(k+1-d_1) \end{bmatrix}^T \tilde{P}^j \begin{bmatrix} Z(k+1-d_1) \\ -\tilde{K}^j Z(k+1-d_1) \end{bmatrix}.
\end{aligned}$$

Then (10.35) will be rewritten as

$$\theta(k) \cdot vec(\tilde{P}^{j+1}) = \phi^j(k). \tag{10.36}$$

Similarly, s groups of data will be collected to obtain $\tilde{P}^{j+1}$, and the initial time $k_0 \geq n + q + d_1$,

$$\Theta \cdot vec(\tilde{P}^{j+1}) = \Phi^j, \tag{10.37}$$

Algorithm 10.2 Online VI Algorithm Based on Measurement Data

1: Give any initial control policy $\hat{u}(k) = u(k) + \hat{e}(k)$ as the input on the time interval $[k_0, k_0 + s]$, where $\hat{e}(k)$ is the exploration noise. Compute Θ and $Z(k)$ in this interval;
2: Select a constant $\epsilon > 0$. Let the iteration index $j = 0$;
3: **Policy evaluation**: solve for $\tilde{P}^{j+1}$ from (10.37);
4: **Policy improvement**: solve for $\tilde{K}^{j+1}$ from (10.38);
5: Repeat Step 3 and step 4 until $\|\tilde{P}^j - \tilde{P}^{j-1}\| < \epsilon$.

where $\Theta = [\theta^T(k_0 - d_1), \ldots, \theta^T(k_0 + s - d_1)]^T$, $\Phi^j = [(\phi^j(k_0))^T, \ldots, (\phi^j(k_0 + s))^T]^T$. Then, the controller gain can be updated by

$$\tilde{K}^{j+1} = (\tilde{P}_{uu}^{j+1})^{-1} \tilde{P}_{uZ}^{j+1}. \tag{10.38}$$

Then, the online VI algorithm, utilizing measurement feedback, is summarized in Algorithm 10.2. The key differences between Algorithms 10.1 and 10.2 lie in the requirement for an initial control policy and the convergence rate of the algorithm. If the initial control policy is admissible, Algorithm 10.1 can be chosen and its convergence rate is relatively faster than that of Algorithm 10.2. If the initial control policy is arbitrary, Algorithm 10.2 can be chosen, but the convergence rate of the Algorithm 10.2 will be relatively slower.

Remark 10.4 It is known from (10.26) and (10.35) that $\tilde{P} \in \Re^{n_{\tilde{P}} \times n_{\tilde{P}}} (n_{\tilde{P}} = 2(n + q) + d_1 + d_2 + 1)$ is a symmetric matrix. To guarantee the application of the least square method, more than $n_{\tilde{P}}^2/2$ sample data should be collected before the learning process in Algorithms 10.1 and 10.2.

Remark 10.5 As usual, a probing noise is added into the input in the process of collecting data, whose purpose is that the system can be excited sufficiently and the data can be collected diversely. Furthermore, the rank condition of matrices Ψ^j and Θ can be satisfied easily. Notice that this condition is necessary when $\tilde{P}^j$ is solved uniquely from (10.33) and (10.37) at each iteration.

10.3.2 Convergence and Tracking Performance Analysis

In this subsection, a rigorous proof is presented for the convergence of the algorithms above, and the tracking performance is analyzed after an approximately optimal controller is applied to the system.

Theorem 10.1 *The sequences $\{\tilde{P}^j\}$ and $\{\tilde{K}^j\}$ obtained by Algorithm 10.1 or Algorithm 10.2 will converge to $\tilde{P}^*$ and $\tilde{K}^*$ as $j \to \infty$, where*

$$\tilde{P}^* = \begin{bmatrix} \tilde{P}_{ZZ}^* & \tilde{P}_{Zu}^* \\ \tilde{P}_{uZ}^* & \tilde{P}_{uu}^* \end{bmatrix}, \tilde{K}^* = (\tilde{P}_{uu}^*)^{-1} \tilde{P}_{uZ}^*$$

with

$$\begin{aligned}\tilde{P}^*_{ZZ} =& M_{2\bar{x}} C^T Q C M_{2\bar{x}} + M_{1\bar{x}} A^T P^* A M_{1\bar{x}} \\ &- M_{1\bar{x}}^T A^T P^* B U M_{1r} - M_{1r}^T U^T B^T P^* A M_{1\bar{x}} \\ &+ M_{1r}^T U^T (R + B^T P^* B) U M_{1r}, \\ \tilde{P}^*_{Zu} =& M_{1\bar{x}}^T A^T P^* B - M_{1r}^T U^T (R + B^T P^* B), \\ \tilde{P}^*_{uZ} =& B^T P^* A M_{1\bar{x}} - (R + B^T P^* B) U M_{1r}, \\ \tilde{P}^*_{uu} =& R + B^T P^* B.\end{aligned}$$

Proof Firstly, consider the convergence of Algorithm 10.1. The PI algorithm for solving the discrete-time ARE (10.30) has been presented in some literature such as [7–9], and the convergence was presented in [24]. Particularly, P^j was solved from

$$P^j = Q_1 + K^{jT} R K^j + (A - BK^j)^T P^j (A - BK^j) \tag{10.39}$$

and sequence $\{P^j\}$ has the property $P^* \leq P^{j+1} \leq P^j$ and $\lim_{j\to\infty} P^j = P^*$. From Lemma 10.1, we have $J^j(k) = \bar{x}^T(k) P^j \bar{x}(k)$, and then $J^*(k) \leq J^{j+1}(k) \leq J^j(k)$, which means that $J^j(k)$ is a non-increasing function with the iteration index j. Meanwhile, the controller gain K^j is obtained from P^j, i.e., $\lim_{j\to\infty} K^j = K^*$. On the other hand, (10.26) is equivalent to

$$P^j = Q_2 + K^{jT} R K^j + (A - BK^j)^T P^j (A - BK^j), \tag{10.40}$$

where $Q_2 = M_{1\bar{x}}^{+T} M_{2\bar{x}}^T C^T Q C M_{2\bar{x}} M_{1\bar{x}}^+$, while $M_{1\bar{x}}^+$ is the generalized right inverse of $M_{1\bar{x}}$ with $M_{1\bar{x}}^+ M_{1\bar{x}} = \mathbf{I}$. Obviously, the sequence $\{P^j\}$ obtained by (10.40) has the same convergence with (10.39). Furthermore, the full rank of Ψ^j ensures the solution of $\tilde{P}^j$ to be unique, which means that P^j and $\tilde{P}^j$ are one-to-one, and $\lim_{j\to\infty} \tilde{P}^j = \tilde{P}^*$. Meanwhile, $\tilde{K}^j$ is determined by $\tilde{P}^j$ uniquely. Hence, $\lim_{j\to\infty} \tilde{K}^j = \tilde{K}^*$.

Secondly, the convergence proof of Algorithm 10.2 is similar to the one of Algorithm 10.1 except the property of P^j. The convergence of VI algorithm for solving the discrete-time ARE was presented in [25], where P^j has the property $P^* \geq P^{j+1} \geq P^j$ and $\lim_{j\to\infty} P^j = P^*$. It implies that the value function $J^j(k)$ is non-decreasing with the iteration index j, i.e., $J^*(k) \geq J^{j+1}(k) \geq J^j(k)$. The following proof is similar to that of Algorithm 10.1, and thus the detailed derivation is omitted here. This completes the proof. □

Based on the convergent algorithms, the next theorem illustrates that the approximately optimal controllers obtained by Algorithms 10.1 and 10.2 can make the system achieve asymptotic tracking, even in the case of external disturbance and time-delay attacks.

Theorem 10.2 *The measurement feedback control policy* $u(k - d_1) = -\tilde{K}^{j*} Z(k - d_1)$ *can stabilize the system (10.15) and guarantee* $\lim_{k\to\infty} e(k) = 0$.

Proof The measurement feedback control policy $u(k-d_1) = -\tilde{K}^{j*}Z(k-d_1)$ is equivalent to $u(k-d_1) = -K^{j*}\bar{x}(k) + Ur(k)$ for all $k > k_0$, and the closed-loop system with the approximately optimal controller becomes

$$\bar{x}(k+1) = (A - BK^{j*})\bar{x}(k).$$

Since the sequence $\{K^j\}$ is convergent and the optimal controller gain K^* can stabilize the system, i.e., $A - BK^*$ is Schur. It implies that there is a small constant $\epsilon > 0$, and a neighborhood with K^* being the center $\delta_\epsilon(K^*)$. Then given any approximately optimal controller gain $K^{j*} \in \delta_\epsilon(K^*)$, we have that $A - BK^{j*}$ is Schur. Hence, $\lim_{k\to\infty} \bar{x}(k) = 0$, which implies $\lim_{k\to\infty} e(k) = 0$. □

Remark 10.6 In Algorithms 10.1 and 10.2, the initial control policies are utilized to generate system trajectories during the data collection period. In the learning phase, the estimation policy $u^j(k) = -\tilde{K}^j Z(k)$ is updated based on the collected data until a certain level of iteration precision is achieved. As a result, Algorithms 10.1 and 10.2 are considered as off-policy learning methods.

Remark 10.7 The proposed algorithms can obtain a measurement feedback controller for any attack-induced delays, provided that sufficient historical data is collected and related conditions are fulfilled. In most practical systems, the delay in the system cannot be too large due to the constraint of attack cost. If the delay exceeds a certain threshold, the information is typically discarded, resulting in a DoS attack. This aspect will be explored further in our future research.

10.4 Simulation

In this section, the mass-spring-damper system (Fig. 10.2) in [23, 26] is used to validate the effectiveness of the proposed algorithms. In Fig. 10.2, M, L and c represent the mass of the object, the stiffness constant of the spring and the damping, respectively. Then, the dynamical process of the mass-spring-damper system is modeled by

Fig. 10.2 Mass-spring-damper system

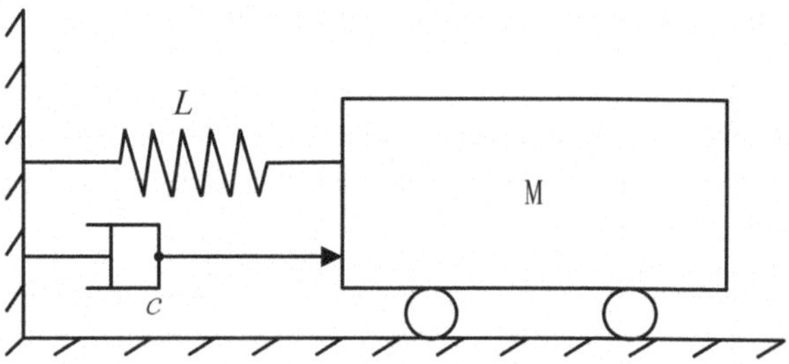

$$\dot{x_1}(t) = x_2(t),$$
$$\dot{x_2}(t) = -\frac{L}{M}x_1(t) - \frac{c}{M}x_2(t) + \frac{1}{M}u(t),$$

where $x_1(t)$ and $x_2(t)$ denote the position and velocity, respectively. Define $x(t) = (x_1^T(t), x_2^T(t))^T$. Then the system can be rewritten as

$$\dot{x}(t) = \begin{bmatrix} 1 & 1 \\ -\frac{L}{M} & -\frac{c}{M} \end{bmatrix} x(t) + \begin{bmatrix} 1 \\ \frac{1}{M} \end{bmatrix} u(t). \tag{10.41}$$

Set the values of the parameters as $M = 1kg$, $L = 4N/m$, $c = 0.5N \cdot s/m$, and consider the bounds of delays d_1=2, d_2=1. Discretize the system (10.41) with $T = 0.1s$. Then the error system with disturbance can be expressed by

$$x(k+1) = \begin{bmatrix} 0.9804 & 0.0969 \\ -0.3876 & 0.9319 \end{bmatrix} x(k) + \begin{bmatrix} 0.0049 \\ 0.0969 \end{bmatrix} u(k-2) + \begin{bmatrix} -1 & 1.5 \\ -2 & 0.5 \end{bmatrix} r(k),$$
$$e(k) = [1\ 0]x(k-1) - [-2\ \ -1]r(k).$$

It's worth noticing that the disturbance and the reference signal are generated by the exosystem, where the exostate vector $r(k)$ is given by the following dynamical system:

$$r(k+1) = \begin{bmatrix} 0.8 & -0.6 \\ 0.6 & 0.8 \end{bmatrix} r(k).$$

The performance index is chosen as (10.16) with $R = 1$ and $Q = 100$. The effectiveness of the proposed algorithms will be validated when time-delay attacks occur in different channels.

A. *The Case of Time-Delay Attacks in the S-C Channel*

The time-delay attacks occur in the S-C channel, and there are attack-induced delays in the output. The initial stabilizing controller gain is chosen as

$$\tilde{K}^0 = [109.4199\ -377.6269\ 441.6297\ -175.1728$$
$$0.1116 \quad 0.1310 \quad -0.3920 \quad 0.4493 \quad 0.8877].$$

By implementing Algorithm 10.1, after 25 iterations, and the approximately optimal controller gain is

$$\tilde{K}^{25} = [209.0733\ -621.5990\ 654.7387\ -243.4947$$
$$0.3110 \quad 0.3341 \quad -0.6817 \quad 0.4174 \quad 1.2339].$$

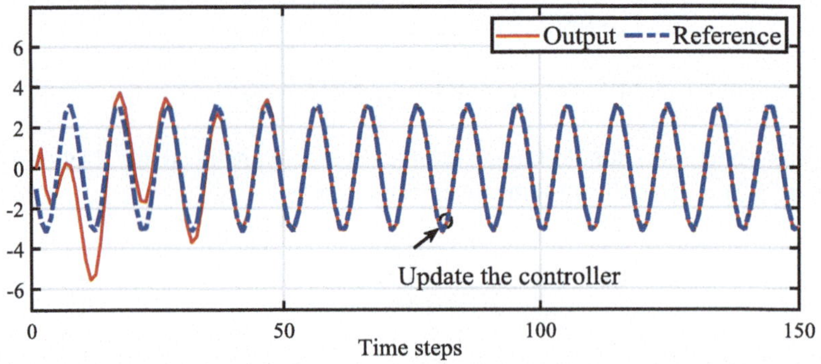

Fig. 10.3 Case of time-delay attacks in the S-C channel

After $k = 80$, $\tilde{K}^{25}$ is applied to the system, the trajectories of the system output and the reference are depicted in Fig. 10.3. It is shown that the output tracks the reference signal asymptotically, which accords with the theoretical results in this chapter.

B. *The Case of Time-Delay Attacks in the C-A Channel*

If time-delay attacks occur in the C-A channel, there will be attack-induced delays in the input. Algorithm 10.2 is implemented, and the initial controller gain is chosen arbitrarily as follows:

$$\begin{aligned}\tilde{K}^0 = [&6.6932\ -106.6709\ 200.6091\ -104.0834\ \ 0.1116\\ &0.1310\quad 0.1443\quad 0.1186\quad 0.5801\quad 0.5274].\end{aligned}$$

The approximately optimal controller gain is obtained after 40 iterations, which is

$$\begin{aligned}\tilde{K}^{40} = [&112.8338\ -393.0753\ 474.5602\ -198.8475\ \ 0.3110\\ &0.3341\quad 0.3430\quad -0.2149\quad 0.6456\quad 1.0076].\end{aligned}$$

Similarly, $\tilde{K}^{40}$ is applied to the system after $k = 80$, and Fig. 10.4 shows that the learned control policy can guarantee the system to achieve the asymptotic tracking. Typically, the variable in VI algorithm requires more iterations than that in PI algorithm to attain the same level of accuracy in a given environment. This is largely due to the different convergence rates of these two algorithms. Notice that a random exploration noise is added to the control input to ensure the persistence of excitation condition during the execution of the algorithms. Here the noise consists of the sinusoids of varying frequencies as shown in Fig. 10.5.

C. *Performance Comparison*

Based on the aforementioned simulations, it is evident that the designed controller successfully achieves adaptive output regulation in the presence of time-delay attacks. In fact, Case A in this chapter was studied in [7], where the delay in one channel was

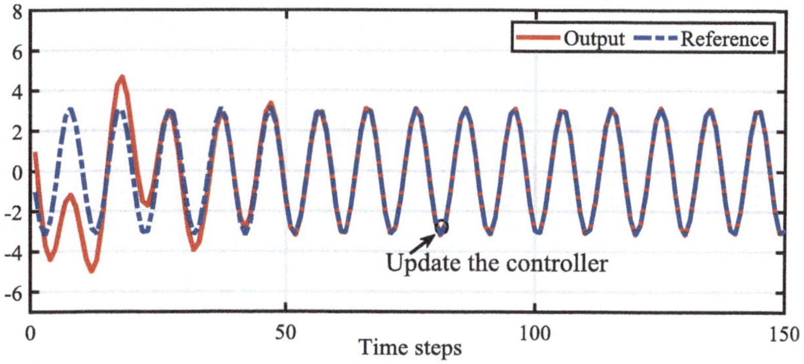

Fig. 10.4 Case of time-delay attacks in the C-A channel.

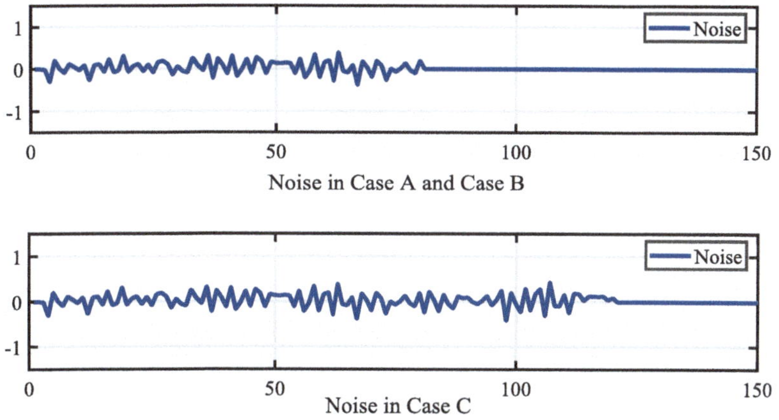

Fig. 10.5 Exploration noises in Cases A, B and C.

addressed. The attack-induced delays in two channels are studied in this chapter, and the maximum delays in different channels may not be the same. If the method of dealing with one-channel delay is applied to the system where two-channels suffer from adversary attacks, the system may be unstable, even be crashed. The simulations also highlighted that the dimensions of the designed controllers differ significantly between the systems affected by one-channel delays and those with two-channel delays. This difference is illustrated in Fig. 10.6. Before updating the controller, the related data is collected along the system trajectory. When the learned measurement feedback controller is applied to a system without accounting for the delay in the S-C channel, the system trajectory deviates significantly from the reference signal. Consequently, the learned controller, specifically designed for the case of input delay, fails to stabilize a system with two-channel delays. When both the C-A channel and the S-C channel are affected by time-delay attacks, the measurement feedback controller introduced in this chapter will be utilized. To evaluate the system's per-

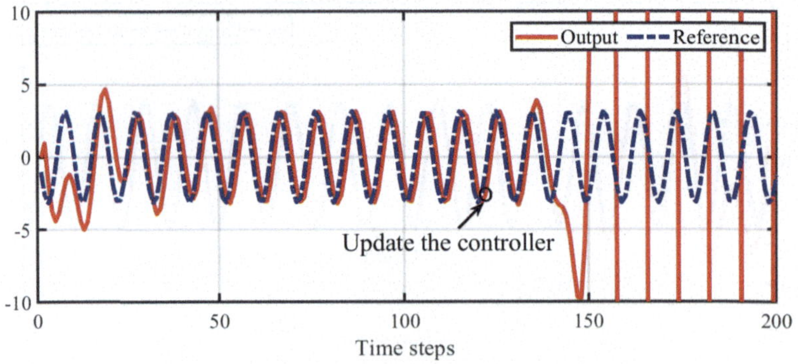

Fig. 10.6 Case of time-delay attacks in two channels with compared method.

formance, Algorithm 10.2 will be employed. An arbitrary initial controller gain will be chosen as follows:

$$\begin{aligned}\tilde{K}^0 = [&-83.1620\ 167.0693\ -81.0967\ -6.3668\ 0.1116\\ &0.1310\quad 0.1443\quad 0.1514\quad 0.5599\ \ 0.5086\ 0.0323].\end{aligned}$$

Then the approximately optimal controller gain is learned after 40 iterations

$$\begin{aligned}\tilde{K}^{40} = [&3.3200\ -91.1307\ 188.9724\ -107.4470\ 0.3110\\ &0.3341\quad 0.3430\quad 0.3382\quad 0.3042\quad 0.6907\ 0.5445].\end{aligned}$$

After $k = 120$, $\tilde{K}^{40}$ is applied to the system, Fig. 10.7 shows the trajectories of the system output and the reference signal. Compared with Fig. 10.6, it is clear that the trajectory changes dramatically after updating the controller. Figure 10.7 shows that the approximately optimal control policy can stabilize the system and make the system output track the reference signal asymptotically, which is as expected for the designed control method in this chapter.

10.5 Conclusion

This chapter studied the adaptive output problem for unknown CPSs under time-delay attacks which occur in the S-C channel and the C-A channel. To address the issue of time-delay attacks and enable real-time closed-loop feedback control, a measurement feedback controller was designed using historical control inputs and tracking errors. The controller's gain was calculated and updated online through an iterative solution of the ARE using the RL method. This approach allows the system to achieve asymptotic tracking and effective disturbance rejection. Simulation results

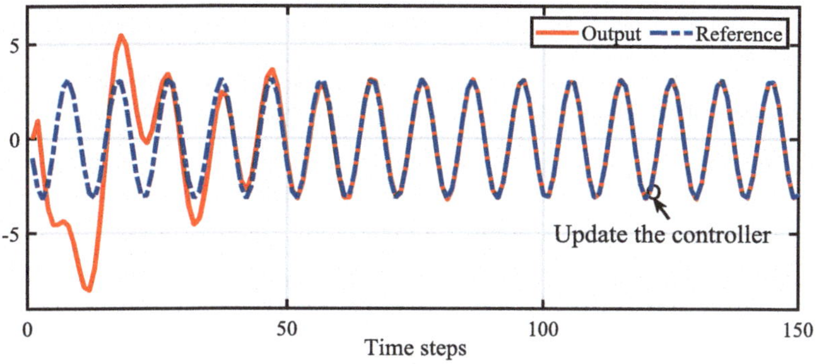

Fig. 10.7 Case of time-delay attacks in two channels with our method

on a mass-spring-damper system confirmed the practical advantages of the proposed control approach. In the future, this method could be extended to cooperative output regulation with set-point attacks and secure control with attack detection.

References

1. Nilsson J, Bernhardsson B, Wittenmark B (1998) Stochastic analysis and control of real-time systems with random time delays. Automatica 34(1):57–64
2. Qi Q, Zhang H (2017) Output feedback control and stabilization for networked control systems with packet losses. IEEE Trans Cybern 47(8):2223–2234
3. Zhang WA, Yu L (2010) A robust control approach to stabilization of networked control systems with short time-varying delays. Acta Autom Sin 36(1):87–91
4. Zhang H, Zhang Z, Wang Z, Shan Q (2015) New results on stability and stabilization of networked control systems with short time-varying delay. IEEE Trans Cybern 46(12):2772–2781
5. Zhang WA, Yu L, Yin S (2011) A switched system approach to H_∞ control of networked control systems with time-varying delays. J Franklin Inst 348(2):165–178
6. Jin D, Yu L, Zhang WA (2017) Stabilization of networked interconnected systems. J Adv Comput Intell Intell Inf 21(2):251–257
7. Gao W, Jiang ZP (2019) Adaptive optimal output regulation of time-delay systems via measurement feedback. IEEE Trans Neural Netw Learn Syst 30(3):938–945
8. Liu Y, Zhang H, Yu R, Qu Q (2018) Data-driven optimal tracking control for discrete-time systems with delays using adaptive dynamic programming. J Franklin Inst 355(13):5649–5666
9. Zhang H, Liu Y, Xiao G, Jiang H (2020) Data-based adaptive dynamic programming for a class of discrete-time systems with multiple delays. IEEE Trans Syst Man Cybern Syst 50(2):432–441
10. Fan J, Wu Q, Jiang Y, Chai T, Lewis FL (2020) Model-free optimal output regulation for linear discrete-time lossy networked control systems. IEEE Trans Syst Man Cybern Syst 50(11):4033–4042
11. Pang ZH, Liu GP, Zhou D, Sun D (2016) Data-based predictive control for networked nonlinear systems with network-induced delay and packet dropout. IEEE Trans Industr Electron 63(2):1249–1257
12. Abouheaf MI, Lewis FL, Mahmoud MS, Mikulski DG (2015) Discrete-time dynamic graphical games: model-free reinforcement learning solution. Control Theory Technol 13(1):55–69

13. Zhao X, Yi P, Li L (2020) Distributed policy evaluation via inexact ADMM in multi-agent reinforcement learning. Control Theory Technol 18(4):362–378
14. Huang X, Dong J (2021) ADP-based robust resilient control of partially unknown nonlinear systems via cooperative interaction design. IEEE Trans Syst Man Cybernet Syst 51(12):7466–7474
15. Sutton RS, Barto AG (2018) Reinforcement learning: an introduction. MIT Press
16. Lewis FL, Vrabie D (2009) Reinforcement learning and adaptive dynamic programming for feedback control. IEEE Circ Syst Mag 9(3):32–50
17. Jiang H, Zhang H, Xie X (2020) Critic-only adaptive dynamic programming algorithms applications to the secure control of cyber-physical systems. ISA Trans 104:138–144
18. Liu R, Hao F, Yu H (2021) Optimal SINR-based DoS attack scheduling for remote state estimation via adaptive dynamic programming approach. IEEE Trans Syst Man Cybern Syst 51(12):7622–7632
19. Huang X, Dong J (2022) Robust dynamic actuator failure compensation control of nonlinear systems via cooperative interaction design. IEEE Trans Cybern 52(7):6013–6023
20. Bradbury M, Leeke M, Jhumka A (2015) A dynamic fake source algorithm for source location privacy in wireless sensor networks. In: 2015 IEEE Trustcom/BigDataSE/ISPA
21. Huang J (2004) Nonlinear output regulation: theory and applications. SIAM, Philadelphia
22. Lewis FL, Vrabie D, Syrmos VL (2012) Optimal control. Wiley
23. Wang D, Mu C (2018) Adaptive-critic-based robust trajectory tracking of uncertain dynamics and its application to a spring Cmass Cdamper system. IEEE Trans Industr Electron 65(1):654–663
24. Hewer G (1971) An iterative technique for the computation of the steady state gains for the discrete optimal regulator. IEEE Trans Autom Control 16(4):382–384
25. Lancaster P, Rodman L (1995) Algebraic Riccati equations. Clarendon Press
26. Modares H, Lewis FL (2014) Optimal tracking control of nonlinear partially-unknown constrained-input systems using integral reinforcement learning. Automatica 50(7):1780–1792

Chapter 11
Attack-Resilient Control Against FDI Attacks Using Sliding Mode Control

11.1 Introduction

The security control problem of CPSs is considered under DoS attacks in Chap. 8 and time-delay attacks in Chaps. 9 and 10. In this chapter, the resilient control problem is studied for CPSs under FDI attacks. Sliding mode control (SMC) is recognized as one of the most effective tools for addressing uncertain systems due to its robustness and insensitivity to perturbations. This robust and effective nonlinear control method has been employed to enhance the security of CPSs. In particular, Li et al. [1] proposed an adaptive SMC law with a discontinuous input term to defend against actuator attacks. Additionally, Cao et al. [2] introduced an SMC method that ensures stochastic finite-time boundedness of CPSs subjected to random injection attacks. It should be noted that the aforementioned studies rely on a common assumption, namely, the system states can be directly measured. However, in practical situations, meeting this stringent requirement is typically challenging. Therefore, observer-based SMC has gained increasing attention. For example, an observer-based event-triggered SMC technique has been proposed for nonlinear networked control systems under the influence of cyber attacks in [3]. In [4], a novel approach based on neural networks was presented to estimate attack signals to mitigate their negative impact, and an adaptive SMC method was designed to ensure the system's security.

This chapter focuses on the control of CPSs in the presence of malicious FDI attacks on actuators. These FDI attacks can lead to actuator faults or even instability of CPSs. To address this issue, we propose an unknown input observer (UIO) to estimate the system states and attack signals. To suppress the impact of FDI attacks, we introduce a discrete-time sliding mode control (DSMC) algorithm. The reaching law of the DSMC algorithm is constructed based on the n-th order difference of the estimation of attack signals. Finally, we present two illustrative examples to demonstrate the effectiveness of the proposed method.

B. Chen et al., *Secure Fusion Estimation and Control for Cyber-Physical Systems Under Malicious Attacks*, https://doi.org/10.1007/978-981-97-9396-9_11

11.2 Problem Statement

Consider a linear discrete-time CPS where the actuator is exposed to FDI attacks. The system's dynamics are written as follows:

$$\begin{cases} x(k+1) = Ax(k) + B\big[u(k) + a(k)\big], \\ y(k) = Cx(k), \end{cases} \tag{11.1}$$

where $x \in \Re^{n_x}$ represents the system state of CPS; $u \in \Re^{l}$ is the control input; $y \in \Re^{n_y}$ denotes the measurement signal; $a \in \Re^{l}$ is the FDI attack signal; A, B and C are constant matrices of appropriate dimensions. T and k denote the sampling period and time step, respectively. For this system, we assume that the pair (A, B) is controllable, and the pair (A, C) is observable.

A robust sliding mode controller can be designed to ensure system stability and effectiveness even after being subjected to attacks. In this design, the corresponding sliding surface s is defined as

$$s(k) = Gx(k), \tag{11.2}$$

where $G \in \Re^{l \times n_x}$ and GB is nonsingular.

To ensure the convergence of system trajectories to the sliding surface in the presence of FDI attacks, it is imperative to design an appropriate UIO and a DSMC law. The objective is that the system trajectories can subsequently converge to the origin along the sliding surface.

11.3 Local Optimal Estimator Design

In this section, a UIO is developed to estimate unidentified FDI attack signals. Additionally, a sliding mode controller is designed to ensure the stability of the control system.

11.3.1 UIO Design

Firstly, the state $x(k)$ can be decoupled into a known input $\bar{x}_1(k)$ and an unknown input $\bar{x}_2(k)$. For this purpose, $\Phi = [M \;\; B]$ and $\Lambda = [CB \;\; N]$ are constructed, where $N \in \Re^{n_y \times (n_y - l)}$, and $M \in \Re^{n_x \times (n_x - l)}$ is a matrix such that Φ is nonsingular. Thus, we have $\Lambda^{-1} = [\Lambda_1 \;\; \Lambda_2]^T$ with $\Lambda_1 \in \Re^{l \times n_y}$ and $\Lambda_2 \in \Re^{(n_y - l) \times n_y}$. Multiplying Φ^{-1} and Λ^{-1} at the both sides of the state and measurement equations in (11.1) yields

$$\begin{cases} \bar{x}(k+1) = \bar{A}\bar{x}(k) + \bar{B}\big[u(k) + a(k)\big], \\ y(k) = \bar{C}\bar{x}(k), \end{cases} \tag{11.3}$$

$$\begin{cases} \Lambda_1 y(k) = \Lambda_1 CM\bar{x}_1(k) + \bar{x}_2(k), \\ \Lambda_2 y(k) = \Lambda_2 CM\bar{x}_1(k), \end{cases} \tag{11.4}$$

where $x = \Phi\bar{x} = \Phi\left[\bar{x}_1\ \bar{x}_2\right]^T$, $\bar{A} = \Phi^{-1}A\Phi = \begin{bmatrix} \bar{A}_{11} & \bar{A}_{12} \\ \bar{A}_{21} & \bar{A}_{22} \end{bmatrix}$, $\bar{B} = \Phi^{-1}B = [0\ \ I_l]^T$, and $\bar{C} = C\Phi = [CM\ \ CB]$. Combining (11.3) and (11.4) yields

$$\begin{cases} \bar{x}_1(k+1) = \tilde{A}\bar{x}_1(k) + Ey(k), \\ \bar{y}(k) = \tilde{C}\bar{x}_1(k), \end{cases} \tag{11.5}$$

where $\tilde{A} = \bar{A}_{11} - \bar{A}_{12}\Lambda_1 CM$, $E = \bar{A}_{12}\Lambda_1$, $\bar{y}(k) = \Lambda_2 y(k)$, and $\tilde{C} = \Lambda_2 CM$. A Luenberger observer can be designed while the system matrices $(\tilde{A}, \tilde{C})$ are observable. According to the concept of UIO, a Luenberger observer is constructed to estimate $\bar{x}_1(k)$ with $\hat{\bar{x}}_1$, which signifies that $\hat{\bar{x}}_1 \to \bar{x}_1$ as $t \to \infty$. The observer is shown as follows:

$$\begin{cases} \hat{\bar{x}}_1(k+1) = (\tilde{A} - L\tilde{C})\hat{\bar{x}}_1(k) + (L\Lambda_2 + E)y(k), \\ \hat{\bar{x}}_2(k+1) = \Lambda_1 y(k+1) - \Lambda_1 CM\hat{\bar{x}}_1(k+1), \end{cases} \tag{11.6}$$

where $L \in \Re^{(n_x-l)\times(n_y-l)}$. Subsequently, the estimated state can be obtained as $\hat{x}(k) = \Phi\hat{\bar{x}}(k)$. Consequently, by substituting (11.6) into (11.3), the FDI attack signal $a(k)$ is estimated as

$$\begin{aligned} \hat{a}(k) =& \Lambda_1 y(k+1) - \Lambda_1 CM\hat{\bar{x}}_1(k+1) - \bar{A}_{21}\hat{\bar{x}}_1(k) \\ &- \bar{A}_{22}\Lambda_1 y(k) + \bar{A}_{22}\Lambda_1 CM\hat{\bar{x}}_1(k) - u(k). \end{aligned} \tag{11.7}$$

Rearranging (11.7), a simplified equation is obtained as

$$\hat{a}(k) = K_1 y(k+1) + K_2\hat{\bar{x}}_1(k) + K_3 y(k) + K_4 u(k), \tag{11.8}$$

where $K_1 = \Lambda_1, K_2 = \Lambda_1 CML\Lambda_2 CM - \Lambda_1 CM\bar{A}_{11} + \Lambda_1 CM\bar{A}_{12}\Lambda_1 CM - \bar{A}_{21} + \bar{A}_{22}\Lambda_1 CM, K_3 = -\Lambda_1 CML\Lambda_2 - \Lambda_1 CM\bar{A}_{12}\Lambda_1 - \bar{A}_{22}\Lambda_1$, and $K_4 = -I_l$. The estimated state $\hat{x}(k)$ and the estimated FDI attack signal $\hat{a}(k)$ will be utilized in the compensating procedure of the subsequent SMC design.

11.3.2 SMC Design

For the design of the SMC, the initial procedure involves constructing a suitable sliding surface, which has been successfully achieved as demonstrated in Eq. (11.2). It is crucial to not only have an appropriate sliding variable but also to establish an effective reaching law. In this particular study, we have adopted a reaching law that can be expressed as follows:

$$s(k+1) = (1 - qT)s(k) - \lambda \text{sign}\big[s(k)\big] + \nabla^n a(k), \tag{11.9}$$

where $\nabla a(k) = CB\hat{a}(k)$; $\nabla^n a(k) = \nabla^{n-1} a(k) - \nabla^{n-1} f(k-1)$; $sign\big[s(k)\big] \triangleq \big[\text{sign } s_1(k), \ldots, \text{sign } s_l(k)\big]^T$; q is chosen to satisfy $q > 0$ and $0 < 1 - qT < 1$; $\lambda > 0$ and $n > 0$ are control parameters to be selected.

The following lemma is given for the FDI attack signal $a(k)$.

Lemma 11.1: $a(k) = O(T), a(k) - a(k-1) = O(T^2)$, and $a(k) - 2a(k-1) + a(k-2) = O(T^3)$, where T is the sampling time.

From Lemma 11.1, it can be further concluded that $\nabla a(k) = O(T)$, $\nabla^2 a(k) = O(T^2)$, and $\nabla^n a(k) = O(T^n)$. Thus, the upper bound of the disturbance term in (11.9) can be decreased by augmenting the order of $\nabla a(k)$. Subsequent analysis will further prove that the width of QSMD can be reduced.

Combining (11.1), (11.2) and the reaching law (11.9), a DSMC law is obtained. Consequently, the corresponding control input $u(k)$ can be mathematically represented as follows:

$$u(k) = -(GB)^{-1}\left[GAx(k) - (1 - qT)s(k) + \lambda \text{sign}[s(k)] + \sum_{i=1}^{n-1} \nabla^i a(k-1)\right]. \tag{11.10}$$

Remark 11.1 As the order of $\nabla a(k)$ increases, there is a reduction in the width of QSMD. However, this also implies a higher constraint requirement for unknown signals. Evidently, the working principle of the proposed DSMC law relies on the statistical characteristics of unknown signals.

11.3.3 Stability Analysis

To date, a DSMC law has been developed for the CPS in the presence of unknown FDI attacks. However, it is crucial to establish the stability of the control system. More specifically, in our scenario, we aim to demonstrate that the switch function $s(k)$ can attain the QSMD from any initial state and sustain it thereafter.

Theorem 11.1 *For the discrete-time CPS as shown in* (11.1) *with*

$$\xi = \max \nabla^n a(k) \leq \lambda, \tag{11.11}$$

the system trajectories will converge to an $O(T^n)$ *QSMD by the proposed DSMC law, where the expression of* Ω_Δ *is shown as below:*

$$\Omega_\Delta = \left\{ s_i(k) \middle| |s_i(k)| < \lambda + \xi \right\}. \tag{11.12}$$

Proof Two cases with $s_i(k) > 0$ and $s_i(k) < 0$ are discussed respectively. For the case with $s_i(k) > 0$, from (11.9) we can obtain

$$\begin{aligned} s_i(k+1) &= (1 - qT)s_i(k) - \lambda + \nabla^n a_i(k) \\ &< s_i(k) - \lambda + \nabla^n a_i(k) \\ &< s_i(k). \end{aligned} \tag{11.13}$$

As shown in (11.13), when $s_i(k) > 0$, the sliding variable $s_i(k)$ is monotonically decreasing. Therefore, there exists a situation where $s_i(k^*) > 0$ and $s_i(k^* + 1) < 0$. It is of necessity to investigate whether $s_i(k^* + 1)$ will exceed Ω_Δ in this case, i.e., whether $s_i(k^* + 1) - (-\lambda - \xi) > 0$. The deduction is expressed as follows:

$$\begin{aligned} & s_i(k^* + 1) - (-\lambda - \xi) \\ &= (1 - qT)s_i(k^*) - \lambda + \nabla^n a_i(k^*) + \lambda + \xi \\ &= (1 - qT)s_i(k^*) + \nabla^n a_i(k^*) + \xi \\ &> (1 - qT)s_i(k^*) > 0. \end{aligned} \tag{11.14}$$

It can be seen that when $s_i(k) > 0$, the sliding variable s_i will converge into Ω_Δ.

For the case with $s_i(k) < 0$, we can prove in the same way. The proof of Theorem 11.1 is thus accomplished. □

Although the convergence of the switch function $s(k)$ to the QSMD has been confirmed for any initial state, it is imperative to prove that the trajectories will remain within the confines of the QSMD rather than escaping from it.

Theorem 11.2 *For the system* (11.1) *with the assumption* (11.11) *and under the DSMC law* (11.10), *when the switch function* $s(k)$ *enters QSMD* Ω_Δ, *it will always stay in it.*

Proof When $s(k)$ enters the QSMD, there exist two situations, namely, $0 \leq s(k) \leq \lambda + \xi$ and $-\lambda - \xi \leq s(k) \leq 0$. Hence, we discuss this issue in two cases.

For the case with $0 \leq s_i(k) \leq \lambda + \xi$, it can be derived that

$$\begin{aligned} s_i(k+1) &= (1 - qT)s_i(k) - \lambda + \nabla^n a_i(k) \\ &\geq -\lambda + \nabla^n a_i(k) \\ &\geq -(\lambda + \xi) \end{aligned} \tag{11.15}$$

and

$$\begin{aligned} s_i(k+1) &= (1 - qT)s_i(k) - \lambda + \nabla^n a_i(k) \\ &\le \lambda + \xi - (\lambda - \nabla^n a_i(k)) \\ &\le \lambda + \xi. \end{aligned} \tag{11.16}$$

For the case with $-\lambda - \xi \le s_i(k) \le 0$, the proof can be completed by a similar procedure.

As a result, we can conclude that once the sliding variable $s(k)$ enters the QSMD Ω_Δ, it remains confined within Ω_Δ and cannot break free from its boundaries.

Thus, the proof of Theorem 11.2 is accomplished. □

11.3.4 Control Parameter Selection

Selection of λ: The parameter λ is responsible for determining the breadth of the QSMD as well as the system's capacity to withstand attacks. When λ is increased, it enhances the system's ability to resist attacks. Nevertheless, this improvement comes at the expense of introducing more chattering. Taking into account this tradeoff, we set $\lambda = 0.2$.

Selection of q: The parameter q plays a crucial role in determining the convergence rate of the system. A higher value of q accelerates the convergence rate, albeit increasing the likelihood of instability. To strike a balance between convergence speed and stability, we set the parameter q as $q = 5$.

Selection of n: The parameter n represents the order of the attack signal in our study. As evidenced by Lemma 11.1, a higher value of n results in a narrower QSMD. Nevertheless, an increase in n is accompanied by more stringent conditions that need to be met. In light of this tradeoff, we set $n = 5$.

11.4 Simulation

A fourth-order inverted pendulum system [5] is adopted as the controlled object whose dynamic model is as (11.1), where the system matrix, input matrix and measurement matrix are set as $A = \begin{bmatrix} 1 & 0.0991 & 0.0136 & 0.0004 \\ 0 & 0.9818 & 0.2789 & 0.0136 \\ 0 & -0.0023 & 1.1598 & 0.1053 \\ 0 & -0.0474 & 3.2764 & 1.1598 \end{bmatrix}$, $B = \begin{bmatrix} 0.0091 \\ 0.1822 \\ 0.0232 \\ 0.4743 \end{bmatrix}$ and $C = \begin{bmatrix} 1 & 0 & 0 & 0 \\ 0 & 0 & 1 & 0 \\ 0 & 0 & 0 & 1 \end{bmatrix}$, respectively.

Two simulations are conducted to evaluate the effectiveness and robustness of the UIO-based SMC law using different FDI attack signals. In Case 1, the FDI attack signal $a(k)$ is set as follows:

$$\begin{cases} a(k) = -2\cos(k\pi T) & k < 50, \\ a(k) = 2 & 50 \leq k < 125, \\ a(k) = 5\sin(k\pi T) & 125 \leq k < 200. \end{cases} \tag{11.17}$$

In Case 2, the FDI attack signal $a(k)$ is designed as

$$\begin{cases} a(k) = \sin(k\pi T) & k < 50, \\ a(k) = \sin(k\pi T) + e^{1-\frac{1}{[(k-50)T]^2}} & 50 \leq k < 125, \\ a(k) = \sin(k\pi T) + e^{1-\frac{1}{[(k-50)T]^2}} - e^{1-\frac{1}{[(k-125)T]^2}} & 125 \leq k < 200. \end{cases} \tag{11.18}$$

By implementing the proposed algorithm, we obtained simulation results, which are shown in Figs. 11.1, 11.2, 11.3, and 11.4. Specifically, the upper parts of Figs. 11.1 and 11.3 display the estimation performance of the proposed UIO. The estimated FDI attack signals accurately track the actual signals under different attacks. Figure 11.2 illustrates that when the attack signal changes (i.e., when $k = 125$), the sliding variable deviates from the QSMD for a short time but quickly converges back to it. The width of the QSMD is approximately 0.2. From Fig. 11.4, we observe that the proposed DSMC law guarantees the convergence within the QSMD, where the width of the QSMD is around 0.17. The lower parts of Figs. 11.1 and 11.3 demonstrate that the system trajectories, under the influence of the presented method, converge to zero rapidly and remain stable thereafter.

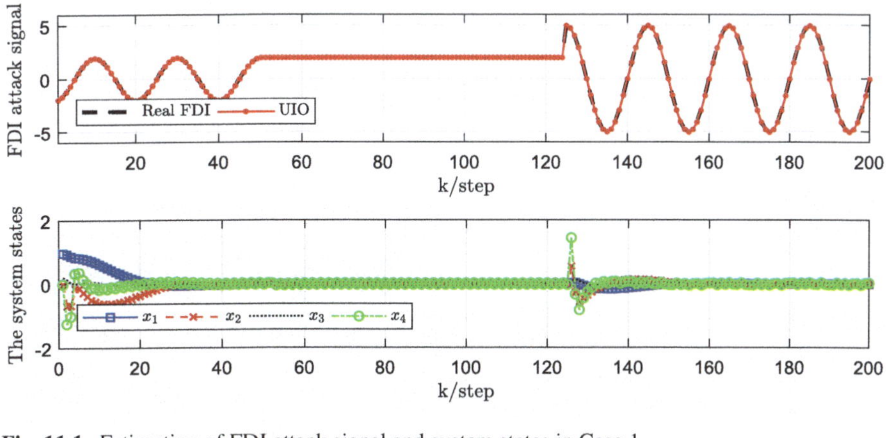

Fig. 11.1 Estimation of FDI attack signal and system states in Case 1

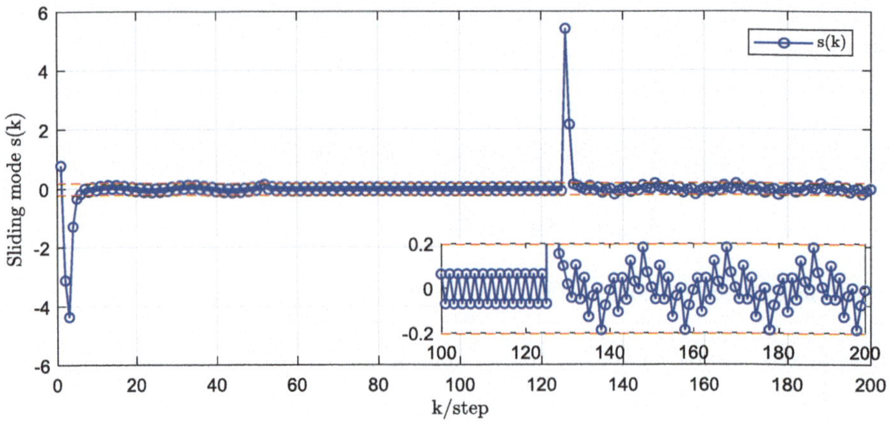

Fig. 11.2 Sliding mode dynamics in Case 1

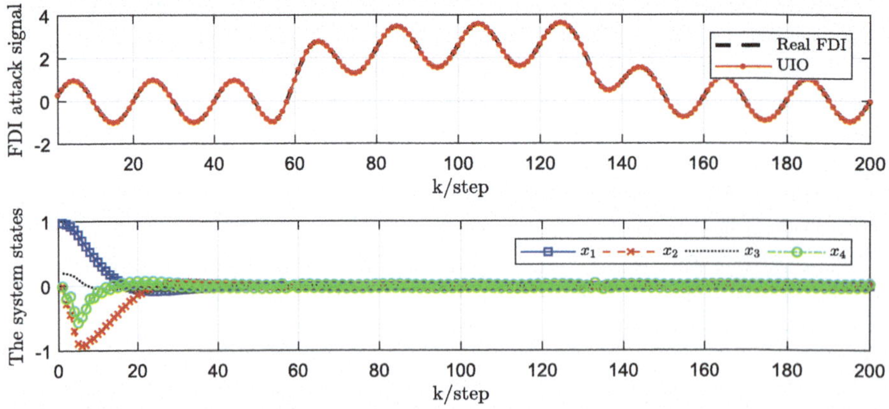

Fig. 11.3 Estimation of FDI attack signal and system states in Case 2

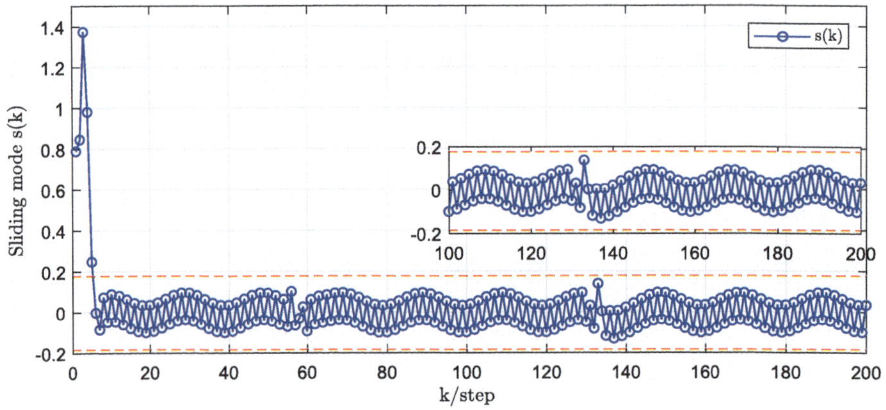

Fig. 11.4 Sliding mode dynamics in Case 2

11.5 Conclusion

This chapter studies the control of CPSs under FDI attacks. A UIO was designed to estimate the system states and detect attack signals. Additionally, a DSMC algorithm was developed to mitigate the impact of FDI attacks. The stability of the control system was mathematically substantiated in detail. The simulation results clearly demonstrate the effectiveness and superiority of the UIO-based DSMC method.

References

1. Li M, Chen Y, Zhang Y, Liu Y (2022) Adaptive sliding-mode tracking control of networked control systems with false data injection attacks. Inf Sci 585:194–208
2. Cao Z, Niu Y, Song J (2019) Finite-time sliding-mode control of Markovian jump cyber-physical systems against randomly occurring injection attacks. IEEE Trans Autom Control 65(3):1264–1271
3. Zhang L, Guo G (2021) Observer-based adaptive event-triggered sliding mode control of saturated nonlinear networked systems with cyber-attacks. Inf Sci 543:180–201
4. Cao Z, Niu Y, Zou Y (2021) Adaptive neural sliding mode control for singular semi-Markovian jump systems against actuator attacks. IEEE Trans Syst Man Cybernet Syst 51(3):1523–1533
5. Yang H, Xu Y, Zhang J (2017) Event-driven control for networked control systems with quantization and Markov packet losses. IEEE Trans Cybernet 47(8):2235–2243

The manufacturer's authorised representative in the EU is Springer
Nature Customer Service Centre GmbH, Europaplatz 3, 69115 Heidelberg,
Germany. If you have any concerns regarding our products, please
contact ProductSafety@springernature.com

Printed and bound by CPI Group (UK) Ltd, Croydon, CR0 4YY
08/10/2026
02229867-0007